D0209716

OXYGEN

Science Essentials

Books in the *SCIENCE ESSENTIALS* series bring cutting-edge science to a general audience. The series provides the foundation for a better understanding of the scientific and technical advances changing our world.

In each volume, a prominent scientist—chosen by an advisory board of National Academy of Science members—conveys in clear prose the fundamental knowledge underlying a rapidly evolving field of scientific endeavor.

O X Y G E N

A Four Billion Year History

Donald Eugene Canfield

Princeton University Press
Princeton and Oxford

Library of Congress Cataloging-in-Publication Data
Canfield, Donald E.
Oxygen : a four billion year history / Donald Eugene Canfield.
pages cm. — (Science essentials)
Summary: "The air we breathe is twenty-one percent oxygen, an
amount higher than on any other known world. While we may take
our air for granted, Earth was not always an oxygenated planet. How
did it become this way? Oxygen is the most current account of the
history of atmospheric oxygen on Earth"— Provided by publisher.
ISBN 978-0-691-14502-0 (hardback)
1. Oxygen. I. Title.
QD181.O1.C36 2014
551.51'12—dc23   2013024610

British Library Cataloging-in-Publication Data is available

This book has been composed in Baskerville

Printed on acid-free paper. ∞

Printed in the United States of America

1 3 5 7 9 10 8 6 4 2

CONTENTS

This book is dedicated to the memory of my father,
Eugene David Canfield Jr., my guiding light.

ACKNOWLEDGMENTS

I must begin by acknowledging all of my good friends and colleagues who have worked hard in various ways to help unravel the dynamics of oxygen cycling on both the modern and the ancient Earth. This book is as much their story as it is mine. You will meet most of these people as the story unfolds, but I would like to highlight the inspiration of Bob Berner, Tim Lenton, Rob Raiswell, John Hayes, Lee Kump, Penny Chisholm, Ed Delong, Nick Butterfield, Jorge Sarmiento, Osvaldo Ulloa, Bo Thamdrup, Bo Barker Jørgensen, Andrey Bekker, Bob Blankenship, Roger Buick, Fritz Widdel, Niels Peter Revsbech, Martin Brasier, Jake Waldbauer, Jochen Brochs, Birger Rasmussen, Bill Schopf, Paul Falkowski, Bill Martin, Dave Des Marais, John Waterbury, Sean Crowe, Simon Poulton, CarriAyne Jones, Jim Kasting, Minik Rosing, Christian Bjerrum, Tim Lyons, Ariel Anbar, Stefan Bengtson, Andy Knoll, Roger Summons, Dave Johnston, James Farquhar, Nick Lane, Jim Gehling, Guy Narbonne, Tais Dahl, Daniel Mills, and Emma Hammarlund. I also wish to acknowledge the constant stimulation of the NordCEE group spread between and the University of Southern Denmark, the University of Copenhagen, and the Swedish Museum of Natural History. Many of the heroes of this story are no longer with us, but their inspiration continues, and these people include Dick Holland, Vladimir Vernadsky, Preston Cloud, Karl Turekian, and Bob Garrels. This book progressed in fits and starts, but I am especially grateful to the Division of Geological and Planetary Sciences at Caltech, and especially to my host Woody

Fischer for arranging a Moore Fellowship to support two months of productive bliss, together with my family and away from the distractions of home. During the course of writing, I received valuable feedback on individual chapters from Bob Blankenship, Minik Rosing, Bob Berner, Tais Dahl, Emma Hammarlund and Guy Narbonne. I am grateful to Bill Martin and to Lee Kump who provided feedback on the whole text, and especially to Raymond Cox, Tim Lyons, and my copyeditor Sheila Ann Dean, whose extensive comments and edits greatly improved the manuscript. I wish to acknowledge both the patience and extensive feedback from my editor, Alison Kalett at Princeton University Press. Images, or the data to generate them, were kindly provided by Minik Rosing, Emma Hammarlund, James Farquhar, Matt Saltzman, Niels Peter Revsbech, Ken Williford, Martin van Kranendonk, Bruce Wilkenson, Bill Schopf, Tais Dahl, Eric Condliffe, Bo Thamdrup, Jakob Zopfi, and Lawrence David. Finally, I wish to acknowledge the generous support from my funding sources including the Danish National Research Foundation (Danmarks Grundforskningsfond), the European Research Council (Oxygen Grant), and the Agouron Institute.

PREFACE

If you are like me, you probably don't think a whole lot about the air you breathe unless, for some reason, it smells bad. However, our air is quite special. It contains 21% oxygen, and ours is the only world we know of (at least so far) with such elevated amounts. This is good for us because we are large animals and we need lots of oxygen to live. So also do our furry friends, cats and dogs, as well as the cows, chickens, sheep, pigs, and other animals on which we base much of our diet. Oxygen burns the fuel that heats our homes, and allows the warm glow of a campfire on a crisp autumn evening. In short, oxygen is a signature feature of Earth; the high levels in our atmosphere define the outlines of our existence, as they also generally define the nature of animal life on Earth.

Given the importance of Earth's oxygen, we might contemplate a series of questions. So, for example, where does all this oxygen come from? Why are the levels so high? What controls the atmospheric concentrations of this important gas? We might further wonder if oxygen concentrations have always been so high and if not, how they have changed through time, and if so, why. Finally, given the importance of oxygen to the present biosphere, is there any indication that the history of atmospheric oxygen levels could be coupled in any way to the history of biological evolution on Earth?

This book is about the history of atmospheric oxygen on Earth, and I will attempt to answer these questions in the following pages. One of the inevitable conclusions, which I offer in advance, is that oxygen control is a global phenomenon, and oxygen persists in high levels because of a fascinating interplay between biological and geological processes. The nature of this interplay has changed through time, resulting in a rich history of oxygen evolution; this history, as well as we understand it, will be revealed in the pages to follow.

The story is also about the people involved in unraveling the history of oxygen evolution. Indeed, understanding this history has become a popular subject, and many scientists are now involved in its exploration. Many of these investigators are good friends and colleagues, and they have all contributed to a wonderful and rich work life. There are also heroes in this story; visionary thinkers who forged the paths down which others, including myself, follow. Some of those thinkers were decades ahead of their time.

This book is also about how we know what we know. I present the evidence. This is based, mostly, on clues left in ancient sedimentary rocks. Some of the evidence is good and some of it is not so good, especially when we look at very old rocks where the ravages of time have taken their toll. The preservation of the geologic record, however, is part of the story, and we must use the evidence that we have. This means that sometimes we are unable to draw firm conclusions. Uncertainty like this is also very much a part of the scientific process, and I therefore draw attention to it. Still, we can often look at a problem with multiple lines of evidence, and if we apply Ockham's Razor,[1] we can usually come to a reasonable working hypothesis as to the meaning of the data. I also try to highlight instances where our ideas have evolved as the data have improved, become more abundant, or are better understood.

Not all of the evidence, however, comes from geology. There is a strong biological component to the story. We sometimes need to look at modern organisms and modern ecosystems to see how they work. They provide important clues to help us understand how the ancient world worked, especially in details that the geologic record can't easily provide. We also must consider biological evolution. How, for example, did biological oxygen production come to be? This is a fascinating story.

Sometimes we also need to understand complex topics of, such as how photosynthesis works, or how isotopes might be used to unravel the history of oxygen. It has been my goal to make these discussions accessible to anyone interested in science, so I try to introduce difficult principles with enough background that they become broadly understandable. I also use endnotes to explain principles and processes in the detail that a specialist or an especially interested lay person might appreciate. My hope, though, is that the story does not really require one to visit the endnotes, unless the reader wants to learn even more.

Finally, this is a story about time—vast amounts of time. Planet Earth is approximately 4.5 billion years old, which is roughly one-third of the age of the universe. I studied chemistry in college, and my experience with time, at least scientifically, was limited to the hours or days of a chemical reaction. Our whole lives are but a blink of an eye in comparison to the age of Earth. Indeed, thinking about immense tracks of geologic time did not come easily to me. The enormity of geologic time challenges us to imagine how slow processes, like evolution or mountain building, can actually work. I'm now more comfortable with geologic time, and with the time scales of geological and evolutionary processes, but I sympathize with the difficulty of perceiving processes that play out over time scales immensely longer than a human life span. Anyway, the vastness of geologic time was recognized centuries ago, and made famous in the closing remarks of James Hutton's classic 1788 book, *Theory of the Earth*:

> The result, therefore, of our present enquiry is, that we find no vestige of a beginning,—no prospect on an end.

Not long after Hutton's book was published, it became apparent that distinct fossil assemblages could be recognized in certain layers of rocks. This was of practical use for identifying layers that might have economic interest, but it also became obvious that these layers could be divided, subdivided, and dated relative to one another. A key principle in dating was the simple deduction, made in the seventeenth century by the Danish polymath Nicolaus Steno, that a layer of sediment deposited over another is younger than the layer below. This is known as the Law of Superposition.

Major divisions were often described based on either the loss or appearance of distinctive fossil groups, and by correlating from one outcrop to another; these divisions could be recognized from place to place and eventually around the world. Divisions were given names, and as

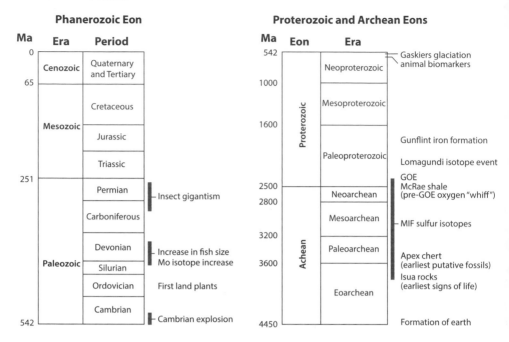

Figure P.1. Geologic time scale showing major events highlighted in the text. Time scale after Gradstein (2004).

radio-isotope dating methods became available, rocks could be dated precisely. What emerges is the geologic time scale. This is our roadmap, our measuring stick, and it is as central to geology as the periodic table is to chemistry. There are multiple scales to these divisions, ranging from eons (hundreds of millions to billions of years), to eras (tens to hundreds of millions of years), to periods (tens of millions of years), and finally to stages (millions of years). An abbreviated version of the geologic time scale is shown in figure P.1, including some key events and places discussed in the text.

Writing this text has been both a joy and an immense learning experience. It has been great fun to try and focus my thinking on topics that have been fuzzy, and to follow the historical development of the many ideas presented here. The only negative aspect of the writing was the quick realization that I could discuss only a fraction of the relevant literature on any given subject. So I apologize in advance to my colleagues and friends whose work has gone unmentioned. It has not gone unnoticed. Despite the need to economize in the text, I hope that this book represents an up-to-date view of the subject, but I recognize that in 30 years, a very different book will likely be written. I hope you enjoy.

Don Canfield (Odense, Denmark)

OXYGEN

CHAPTER 1
What Is It about Planet Earth?

I'm sitting on the train, as I often do, traveling between Odense and Copenhagen. We've just pulled from the stop at Ringsted. I look out the window. The scene is typical Danish countryside of mixed farmland and forest. I pass cows grazing lazily in the field, and beyond them, a farmer is cutting hay. High above, a hawk searches for mice in the uncut grass. I love this landscape. It reminds me of the Ohio countryside where I grew up. Not spectacular, but somehow comforting and reassuring; an honest landscape not prone to bragging or trickery. I squint, and the landscape merges into a mass of green, the cows become ghosts in the distance. I open my eyes again, and we pass a small patch of dense forest (or at least what passes for forest in Denmark). My mind wanders and I reflect on what I see. Denmark is a small country and the land, including the forests, is heavily managed, so the diversity of life isn't terribly high. You could to go the rain forests of Costa Rica or Brazil and be far more impressed with the tropical birds, frogs, insects, and the abundant greenery. Still, even in Denmark, the landscape is brilliant green and teeming with life. Indeed, no matter how you look at it, Earth is defined by abundant and diverse life. The question that preoccupies me now is why?

One might suggest that all the life we see is simply a consequence of a long history of biological evolution on Earth. In his wonderful book *Life on a Young Planet*, my colleague and good friend Andy Knoll from

Harvard University documents the changing face of life during the first four billion years of Earth history. He shows how a variety of biological innovations, like the invention of oxygen-producing photosynthesis, for example, fundamentally shaped the history of life. After oxygen-producing organisms first evolved, other organisms that use oxygen followed, and they then prospered, multiplied, and evolved into yet other oxygen-utilizing life forms. Eventually this led to animals, the most biologically complex of all organisms on Earth. With no oxygen, there would be no animals. So, clearly, innovations during biological evolution have shaped, evened defined, the biosphere. But does evolution alone explain the bounty of life on our planet?

To consider this question, we quickly compare Earth and Mars. Scientists still hold out for the possibility of life on Mars: after all, Mars is the same age as Earth and there is some evidence for at least occasional surface and subsurface water on the planet. Even as I write, NASA's rover *Curiosity* is probing the Martian surface for signs of water, and for clues as to how water interacts with the planet's surface environment. As we will discuss more fully below, and as the tenet goes, where there is water, there may be life. Yet, if there is life on Mars, it doesn't jump up and down like the Whos in Whoville, crying: "We are here, we are here, we are here!" In contrast, if intergalactic explorers probed Earth as we presently probe Mars, it would be impossible to miss Earth's abundant life. The question is, quite simply, why is there so much life on Earth?

To answer this we will for the moment abandon considerations of evolution and start with a more fundamental question: What are the basic ingredients needed for life, at least for life as we know it? As I digest my lunch of lasagna leftovers, I proclaim that food must be important. Yes indeed, but not all organisms can eat lasagna, and I'm reminded of a whole class of creatures who don't eat any kind of organic matter at all, but rather make their cells from simple inorganic substances. Plants fit this bill, growing from carbon dioxide and water and using the energy of the Sun to combine these compounds into cell biomass and oxygen.

Many other types of organisms also fit the bill, and most of them do not use the Sun for energy. Rather, they gain their energy from promoting the reaction between inorganic substances in so-called oxidation-reduction reactions, where electrons are transferred during the reaction.

To probe this idea further, let's think of salt. Put salt in water and it dissolves in a reaction that yields energy, but organisms cannot grow from the energy of this reaction; no electrons are transferred; and the chloride and sodium atoms have the same charge in the salt crystal as they do in the solution. Now think of cows. Cows house enormous populations of microbes in their digestive system, and many of them form methane. Many of these microbes, so-called methanogens, grow quite happily by combining hydrogen gas and carbon dioxide to form methane gas. No light is used, electrons are transferred, the methanogens are happy, and so, presumably, are the cows. Therefore, a basic necessity for life is energy, which is supplied either from light, or from a myriad of different oxidation-reduction reactions.[1] We will look at these issues in more detail in the next chapter, but for now, it's sufficient to highlight that energy is critical for life.

Energy is critical, but we need other things too. Cells are made up of carbon, oxygen, hydrogen, nitrogen, phosphorus, and sulfur as the major ingredients, with a whole suite of trace metals and other elements as well. All of these compounds are critical in the construction of basic cellular components like the cell membrane, genetic material (DNA and RNA), and all of the proteins and other molecules used in running the cell's machinery.

Another basic ingredient of life, at least for life as we know it, is a stable aqueous (meaning water) environment. Life likes it wet! Many organisms, of course, have evolved to live outside of the watery sphere of our planet, but they still all need water to live. So do we, but we just pack it inside our bodies. So, whether we're talking about desert cacti, spiders, snakes, trees, or the smallest bacteria, they all need water. Indeed, this is one reason, as mentioned above, why the search for life in our solar system and beyond is tantamount to searching for liquid water. "Wait," you might say, "I've heard about small bacteria and algae living in sea ice and even in glacial ice in some cases." Very true, but if the organism is alive and growing,[2] it has access to liquid water. In the case of sea ice, this could be brine channels formed as salt is excluded from the growing ice; or for glaciers, high pressure induces ice melting near the bottom, providing an aqueous environment for organisms. "Well then," you might add, "I've heard that the temperature record for a living organism is about 120°C (248°F), well above the boiling point of water at Earth's surface." True again, but these organisms are only found at high pressures, like deep in the ocean where the boiling point of water exceeds the upper temperature limit for life.

What is the big deal about water anyway? For one, water has special properties. Because of its physical structure, a water molecule is bipolar, which means that it is slightly charged with a positive charge on one side and a negative charge on the opposite side. This condition allows it to dissolve all kinds of so-called ionic chemical substances (also charged), many of which constitute the building blocks of life. These include nutrients like nitrate, ammonium, and phosphate, which form into critical components of DNA, RNA, and cell membranes, as well as a host of other substances including sulfate and a variety of trace metals, which help to build the biochemical machinery of the cell. Not only does water dissolve the substances, but these substances are also transported by diffusion and advection; and this movement provides a means by which they can be supplied to the cells. Water also provides the medium by which waste products can be exported from the cell.

The bipolar nature of water also allows for the formation of cell membranes. These separate the external environment from the inside of the cell where the business of life is conducted. Cell membranes are made up of special (phospholipid) molecules with one end containing water-loving chemical groups (hydrophilic) and the other end containing water-repelling chemical groups (hydrophobic). In forming a membrane, the water-loving side reaches out toward the water phase, while the water-repelling side reaches in and lies foot to foot with another row of water repelling bits whose water-loving sides reach out in the opposite direction. This lipid bilayer joins in a circle forming the cell membrane, separating the inside of the cell from the outside environment. All in all, from its ability to dissolved and transport the chemical constituents of life, to its ability to host membrane structures, water is a unique chemical substance.

Or maybe we're thinking too small, too Earthcentrically. Water is the fluid of life because its properties are perfect for the type of life that we know. Perhaps a different type of life could have evolved in different solvents with different properties. It's hard to rule this possibility out. Alternative potential solvents are sometimes named. These include ammonia, methane, sulfuric acid, or hydrogen fluoride (HF); at the right temperature and pressure, they share some (but not all) of the properties of water. Aside from numerous science fiction books and movies, there is also an active scientific literature on this fascinating topic. Discussions of life in these alternative solutions are, however, highly speculative; one might even say imaginative. Therefore, I'll take the easy road, and as far as we know for certain, water is the perfect and only solvent for life.

To summarize, we have highlighted three basic ingredients for life. These are energy, the chemical components that make up cells, and water. We will see that the availability of each of these is linked by special properties of planet Earth.

Let's start with water. It's no secret that Earth is a watery planet. From NASA's spell-binding images of our "blue planet" from space, to the "Rime of the Ancient Mariner" by Samuel Taylor Coleridge, we are reminded of the boundless expanse of the global oceans. We will not concern ourselves at length with why Earth has so much water—likely a combination of early degassing from its interior as well as delivery from comets—but rather with why the water we have is, well, wet. The answer of course is that most of the planet is of the right temperature, lying between the boiling and freezing points of water. But why? Here, at least in part, we are lucky. We can think of it like this. Earth sits at a certain distance from the Sun as dictated by its orbit. The Sun has a certain brightness as dictated by its size and chemical composition.

The amount of the Sun's warmth intercepted by Earth depends on a combination of these two factors. However, as all planets of our solar system are warmed by the same Sun, let's consider distance from the Sun as the key variable. It's easy to imagine that if Earth was closer to the Sun it would receive more warmth, and less warmth if further away. As it turns out, Earth resides at a distance from the Sun where the warmth is sufficient to allow liquid water to persist. If closer to the Sun like Venus, the temperature becomes too hot, and liquid water is boiled away into the atmosphere in a so-called runaway greenhouse. Some of this water may even be completely lost through chemical processes in the stratosphere. If further from the Sun, like Mars, the surface would become too cold and therefore frozen. The zone defining the optimal distance from the Sun (or any other star in fact) where liquid water can persist is known as the "habitable zone," which is sometimes referred to as the "Goldilocks Zone."[3]

But distance from the sun is only part of the story. Earth has an atmosphere with greenhouse gases that contribute to surface warming. Without any greenhouse warming, and with surface albedo as it is,[4] Earth would be frozen with a temperature of −15°C (5°F) or so. Therefore, discovering the habitable zone of a planet is more involved than described above. This requires some rather complex heat-budget calculations, which were first attempted decades ago; however, the most widely referred to models were presented in 1993 by Jim Kasting of Penn State University, along with his coworkers Daniel Whitmire and Ray Reynolds. Jim has been a leader in applying his detailed knowledge of atmospheric chemical dynamics to understanding the evolution of both Earth's atmosphere and atmospheres beyond our own. To attack the habitable zone issue Jim tried, through his model, to keep liquid water on the planet by changing atmospheric CO_2 (carbon dioxide) levels, as these control greenhouse warming. One can easily imagine that different atmospheric CO_2 levels would be needed to maintain a habitable zone in response to differences in solar luminosity, which is basically the intensity of a star; and differences in solar luminosity are expected as one travels either away from or toward the star, or in our case, the Sun.

With Jim's model, the outer reaches of the habitable zone are encountered when atmospheric CO_2 concentrations become so high that CO_2 clouds form. These clouds block solar radiation from reaching the planet surface and thereby increase planetary albedo. The end result is a frozen planet. There were other considerations in Jim's modeling that I won't get into here, but in the end, Jim and colleagues concluded that Mars probably lies just outside of the habitable zone. Likewise, Venus also lies outside of the habitable zone. In this case, solar luminosity is simply too high. Even with miniscule levels of atmospheric CO_2 supplying minimal greenhouse warming, the planet surface becomes so hot that water boils into the atmosphere. This situation generates a runaway greenhouse and very high surface temperatures because water is also a good greenhouse gas (and the most important on the modern Earth!).[5] By some of Jim's calculations, the inner edge of the habitable zone may lie as close as 95% of the distance from the Sun to Earth. This is about

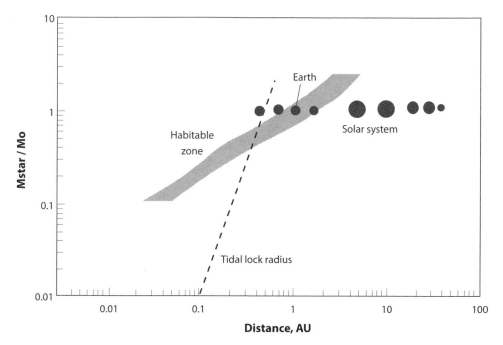

Figure 1.1. Habitability zone as determined by Jim Kasting and colleagues including the placement of the eight planets (plus Pluto) of our Solar System. One AU is one Earth distance from the Sun. The vertical axis shows the ratio between the mass of a star and the mass of the Sun. At distances less than the tidal lock radius from a star, planets become locked in rotations around their axis with small integer values relative to the time scale of the planet's orbit around the star (Mercury rotates 3 times on its axis for every 2 orbits around the Sun). In some cases a planet can orbit with a 1/1 rotation to orbit ratio, with the same side of the planet always facing the star. Planets within the habitable zone of small stars are within the tidal lock radius. From Kasting (2010).

4.5 million miles closer to the Sun than we are. The results of Jim's calculation are presented in figure 1.1, and by all accounts we are lucky; Earth sits snugly within the habitable zone of the Sun.

If this is true, why do we keep entertaining the possibility for life on Mars? Consistent with Jim's habitable zone arguments, there's no evidence for continuously standing surface water on Mars, at least not now. But during decades of satellite and surface exploration, including the recent, highly successful rovers, *Spirit* and *Opportunity* of the Mars Exploration Rover Mission (MER), and the THEMIS (high resolution thermal imaging system) imager onboard the Mars Odyssey orbiter, water has flowed and still does occasionally flow on Mars. This is evidenced by all sorts of channels, ditches, pools, and sedimentary rocks, whose formation is best explained by the action of water. Indeed, the *Curiosity* rover recently landed on the Mars surface and is, as I write, exploring the surroundings of its landing site, which appear to be an ancient river bed! All of this is in addition to spectroscopic observations of water just at and below the soil surface. So, Mars demonstrates that liquid water may be found, at least occasionally, somewhat outside of the habitable zone. By contrast with Earth, however, any life on Mars, if it exists at all, is not obvious and is seemingly restricted in its abundance and occurrence. Therefore, Mars does not and cannot support the magnitude of life that we find on our planet.

Buried in the discussion of Jim Kasting's habitable zone calculations is the idea that over long time scales, Earth actually regulates its own temperature. This idea was first raised by the cosmologist Carl Sagan. Sagan contributed greatly to our understanding of the composition of planetary atmospheres, and he helped frame the discussion about the search for life in the universe. He was an enormous inspiration to those interested in science through his PBS (Public Broadcasting System) program COSMOS, which was originally broadcast in 1980. However, of more importance here, he and his colleague George Mullen asked why Earth didn't freeze early in its history when the Sun was much less luminous than today.[6] Geological evidence points to the more or less continuous presence of liquid water for as far as back as 4.2 billion years ago. Yet, with the present abundance of greenhouse gases in Earth's atmosphere, the planet should have been frozen under the reduced luminosity of the early Sun. This is famously known as "The Faint Young Sun Paradox." Sagan and Mullen argued that this paradox could be solved with a high concentration of greenhouse gases like ammonia and methane; these gases are unstable in our present oxygenated atmosphere but could have been present in the oxygen-poor atmosphere of early Earth. It was soon pointed out, however, that ammonia would be photochemically unstable, even in an oxygen-free atmosphere. This generated a serious problem for the model. However, in a true intellectual quantum leap, Jim Walker, Paul Hays, and Jim Kasting recognized that CO_2 may well have been the greenhouse gas mitigating against an early frozen Earth. Okay, CO_2, big deal. But there is much more to this proposal, because Walker, Hays and Kasting also demonstrated a mechanism that actually regulates surface temperature.

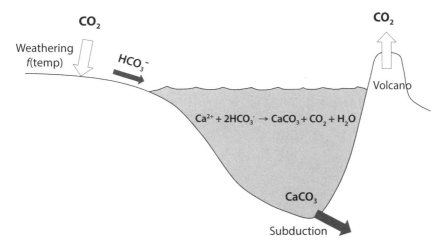

Figure 1.2. The carbon cycle as it acts to regulate Earth's surface temperature. Redrawn from Kasting (2010).

The logic goes like this. Carbon dioxide is constantly introduced from the interior of Earth into the atmosphere. The CO_2 comes from volcanoes and from hydrothermal vents at the bottom of the ocean. However, if we look carefully, we see that these CO_2 sources, at least most of them, originate as a result of Earth's continuous churning in a process known as plate tectonics. In practice, the loss of heat from Earth's interior (estimated at some 5000°C in the middle) causes the mantle (the layer just below Earth's crust) to move and mix in a process known as convection. This convection creates regions of volcanic outpourings, mostly into the oceans, that divide Earth's crust into a series of mobile plates riding on the mantle below. As new ocean floor is formed by this process, old ocean floor is also injected back into the mantle in a process known as subduction (see fig. 1.2). This is a violent process generating most of the major earthquakes, and it is a prime builder of mountain ranges. So, CO_2 is liberated to the atmosphere, but it doesn't accumulate forever. Indeed, it is actively removed by a process known as chemical weathering, where the CO_2 reacts with rocks at Earth's surface.[7] A particularly interesting aspect of weathering is that it is temperature sensitive; it goes faster at higher temperatures.

With this in mind, we can start to imagine how planetary-scale temperature regulation might work. If atmospheric temperature gets too high for some reason, the weathering rate will increase, and CO_2 will be more actively removed from the atmosphere. The increased removal of CO_2 will in turn cause the CO_2 concentration in the atmosphere to drop, reduce the greenhouse warming, and as a result, the temperature will drop. Therefore, a balance point is reached between CO_2 concentration, temperature, and the removal rate of CO_2 by weathering. Suppose for some reason Earth becomes completely frozen. This may have happened a few times during the course of Earth's history. If so, we need not worry, at least when considering long geologic time scales. Tectonic processes ensure that CO_2 will continuously be added to the atmosphere. Without liquid water, there will be no CO_2 removal by weathering, so the CO_2 concentration will build up until temperatures rise to the point where ice melts, and weathering commences again.

During weathering, CO_2 is converted to a soluble ion known as bicarbonate (HCO_3^-), which precipitates as minerals like calcite and dolomite (think of clam shells and coral reefs) in the oceans. These minerals are decomposed back to CO_2 during the subduction processes, thus completing the cycle. To summarize then, Earth, through the cycling of rocks (also known as the rock cycle), has an active control mechanism for temperature, which is enabled by the churnings of the mantle and the associated process of plate tectonics. Therefore, plate tectonics is also critical in allowing Earth to enjoy a continuous record of water through most of its long history.

This is a beautiful story, but is it true? I think that it must be, at least in its broad detail. Some geological evidence, however, points to early-Earth concentrations of atmospheric CO_2 that were too low to warm an Earth illuminated by a less powerful Sun.[8] Jim Kasting has again stepped into the discussion by suggesting that methane may have been, harking back to Sagan and Mullen, a major greenhouse gas early in Earth history. This would help explain the low CO_2 concentrations.[9] This may also be true, but the methane cycle does not obviously lend itself to robust temperature control like CO_2. Very recently, Minik Rosing and colleagues (we meet Minik again in chapter 7) argued that maybe we've been thinking about the problem incorrectly. They suggest, in fact, that maybe the albedo of early Earth was much lower than today,[10] so perhaps we didn't need as much greenhouse gas to warm the planet. Jim Kasting isn't terribly happy with this idea, but lower concentrations of atmospheric CO_2 can satisfy both the geological evidence for ancient CO_2 levels and produce enough of a greenhouse effect to warm the planet in the presence of a faint young Sun. Therefore, the CO_2 control mechanism as originally described by Walker and Kasting can still work to regulate Earth's temperature through time, even if ancient CO_2 levels were lower than we once thought.

Now back to the original question. It's one thing to have water, but it's another thing to support an abundant biosphere. As mentioned in the beginning of this chapter, life is nearly everywhere on Earth's surface. But how does our planet support it? Let's try some calculations. Photosynthetic life on Earth, working at present rates of photosynthesis, would deplete all of the CO_2 in the atmosphere in nine years.[11] Likewise, photosynthetic life in the oceans would deplete all of the available phosphorus, a key nutrient in making aquatic plants and algae, in just 86 years.[12] If this is true, how can we support so much life over long time scales? Part of the answer is that most of the CO_2 and nutrients tied up in plants and algae are liberated back to the environment as these organisms die and are consumed and decomposed by all manner of creatures from giant pandas to bacteria. Okay, but still some plant material and phosphorus aren't liberated back to the environment, and instead these things get buried in sediments and formed into rock. If we redo our calculations based on these rates of loss, we find that CO_2 would be depleted in 13,000 years,[13] and phosphorus in 29,000 years. These are still pretty short time scales compared to the billions of years that life has existed on the planet and the hundreds of millions of years that plants and animals have populated the land surface. How do we explain this?

The answer is actually quite simple. We appeal to the same tectonic processes we used to explain the role of CO_2 in solving the faint young Sun paradox. Luckily, when materials are sequestered into marine sediments on Earth, they are not permanently trapped there. The tectonic movements of the planet ensure that they are not. Through the processes of subduction, mountain building, and sea-level change (sea level is influenced by both tectonics and climate) most of these materials will be exposed again to the weathering environment. During weathering, organic matter is turned back to CO_2, phosphorus is liberated again to solution, and a whole host of other ingredients for life become available once more to support the growth of organisms. The key here is that the magnitude of life we enjoy on Earth is possible because of the active recycling of life's constituents by tectonic processes. This was first recognized over two hundred years ago by James Hutton, whom we also met in the Preface. He wrote the following in his treatise *Theory of the Earth* (1788):

> The end of nature in placing an internal fire or power of heat, and a force of irresistible expansion, in the body of this Earth, is to consolidate the sediment collected at the bottom of the sea, and to form thereof a mass of permanent land above the level of the ocean for the maintenance of plants and animals.

Finally, what about energy? I will say much more about energy in the next chapter, particularly about the types of energy needed for life, many of which you normally wouldn't think about. On modern Earth though, most (probably over 99%) of the energy to the biosphere ultimately comes from the Sun, driving the photosynthesis of plants, algae, and microbes (known as cyanobacteria; we will hear much more about them in later chapters) that produce organic material and oxygen. These products of photosynthesis are biologically recombined in Earth's great food chains. For example, copepods in the ocean eat algae, small fish eat the copepods, larger fish eat the small fish, and even larger fish eat these. These fish die and are decomposed by a variety of bacteria, which in turn are consumed by other organisms. The chain goes on and on but it is fueled, ultimately, by the organic matter and oxygen produced by photosynthesis. As described above, however, the organisms producing the oxygen, and driving the biosphere, obtain their building blocks from material recycled through plate tectonics. Thus, while the Sun supplies the energy, the rates at which tectonics recycles basic biological components sets the tempo.

All in all, we must agree that Earth is a pretty terrific place for life. It sits comfortably within the habitable zone of the Sun. In addition, its active tectonics both control the temperature of the surface environment, providing a continuous supply of liquid water, and recycle the basic components required to fuel abundant life. As we will see in the next chapter, the same tectonics may have also provided optimal conditions for the earliest biosphere.

CHAPTER 2
Life before Oxygen

It was literally the ride of my life. "Do you get claustrophobic?" the man asked. "No, not at all," I lied.[1] "Good," he replied, "and whatever you do, don't touch the red handle. That's used only in emergencies." With a few more instructions, the hatch was closed, and we detached from the crane. We were left to bob freely in the ocean waves, and I waited in anticipation of our descent.

I was sitting in *Alvin*, America's premier deep-diving submersible. With me was my good friend and colleague Bo Barker Jørgensen, now at the University of Aarhus in Denmark, and our pilot Jim. *Alvin* was first commissioned in 1964, and it has been the primary vehicle for deep-sea discoveries in the decades since. A dedicated support crew keeps *Alvin* in tip-top condition, and I was told that probably no parts remain from the original vessel. Still, once inside *Alvin* (at least in 1999), one is reminded of the golden age of space exploration with toggle switches and incandescent indicator lights behind small glass spheres. You feel like you're sitting in the lunar rover, with robust technology that works. What *Alvin* lacks in sophistication, it also lacks in amenities. All three occupants are squeezed into a 2 meter diameter titanium sphere, with the scientists sitting at opposite ends on small foam mats. Necks are cocked to peek out of the two small viewing ports, aiming downward, or alternatively, eyes are fixed on the video monitors above. The body heat of the occupants warms the bathysphere. There is a tank of oxygen to resupply this precious gas, and a cartridge to remove the CO_2 accumulating in the air. Don't ask about the toilet facilities.

We were floating some 1500 meters above the seafloor of the Guaymas Basin in the Gulf of California. Running through the Gulf of California is a spreading zone known as the East Pacific Rise, which separates the North American plate to the east from the Pacific plate to the west. The divergence of these plates slowly drives the Baja Peninsula away from mainland Mexico. This spreading center is somewhat unusual because it includes a thick sediment cover of about one kilometer deposited over millions of years from particles delivered by the once raging Colorado River.[2] Seawater circulates through hot rocks of the spreading center, forming hydrothermal fluids that percolate upward through the sediment and emerge onto the seafloor, precipitating huge mounds of gypsum ($CaSO_4 \cdot 2H_2O$) and supplying abundant sulfide to the local environment.

We begin our descent through the water column. My nose is glued to the observation port as we pass slowly through the upper illuminated zone of the ocean.[3] The light disappears into blackness, and I see the occasional luminescent flash of an unidentified sea creature. Few words are spoken, but none are necessary as we sink through the darkness to the sounds of Brahms playing from the cassette tape player supplied by our pilot. After about one hour, *Alvin*'s outside lights are turned on, and I blink at the most otherworldly sight. Great mounds of *Riftia* tubeworms rise from the shadows,[4] swaying gently on expansive hills of gypsum crust. These elegant sea animals, so beautiful in life, have no mouth or anus and live by cultivating sulfide-oxidizing bacteria in their gut. *Riftia* have evolved an elaborate mechanism for transporting both oxygen and sulfide to the bacteria, which survive by combining these substances. If we look closely, what appears to be gently fallen snow on the gypsum crust is actually a population of free-living sulfide-oxidizing bacteria of the genus *Beggiatoa* (plate 1). They extend to the edge of *Alvin*'s lights. Numerous other animals are also seen, and in one way or another, these are living off the bounty of microbial life supported by the sulfide emanating from the hydrothermal solutions. As a reminder of this, all around us, we see the effervescence of hot, sulfide-rich, hydrothermal waters percolating from the accumulating crust. Sulfide feeds the bacteria, which feed the animals.

The abundant life we find at the Guaymas Basin and many other deep-ocean hydrothermal systems around the world is fueled by sulfide from the vents, but critically, oxygen is also required. These sulfide-oxidizing bacteria live by combining the sulfide with oxygen. Take away the oxygen, and what do we have left? The animals would disappear, the sulfide-oxidizing bacteria couldn't survive, and indeed, almost every sign of life that dominates our view through *Alvin*'s portholes would be absent. How about on a planetary scale? From the last chapter we saw that on today's Earth, over 99% of the energy fueling the biosphere comes from the Sun as channeled through oxygen-producing photosynthesizers. Take away the oxygen producers and all of the food they produce, and the great food chains of Earth would collapse leaving, well, what? This question becomes relevant when we seek to understand the nature of life on ancient Earth, before the evolution of oxygen production.

To answer this, we return to the Guaymas Basin (or nearly any other spreading center with hydrothermal emissions). Take away oxygen, and life would be greatly diminished, but there would still be some to find. Let's think about what the hydrothermal fluids supply. As noted above, in the oxygen-free depths of the oceans, sulfide would be of little use to life, but hydrothermal fluids also contain lots of other compounds, some of which are quite interesting to organisms. We start with hydrogen gas and CO_2, both of which can reach quite high concentrations in hydrothermal vent fluids. Remember the cows from the last chapter? The same type of (autotrophic[5]) methanogens we find in their digestive systems can also combine H_2 (hydrogen) and CO_2 to produce methane in hydrothermal vent systems.[6] Indeed, many methanogens are adapted to very high temperatures of over 100°C and are found associated with modern hydrothermal vents.

Even without oxygen, these methanogens would grow and divide and some would die. The population would reach a steady state of sorts, where growth matched death, and the dead organisms would be food for other microbes. Today, fermenting bacteria play a major role in the decomposition of the organic compounds produced by organisms. They gain energy and grow through the fermentation process whereby they produce simple organic molecules that can be used by other microbes. Indeed, methanogens of a different type (heterotrophic[7]) can use these simple organic compounds and produce methane and CO_2. Therefore, we see the possibility of an ecosystem running on methane, with primary producers consisting of autotrophic methanogens, as well as consumers, which include various fermenting bacteria and heterotrophic methanogens.

This, I think, is a realistic view of the major microbial players populating ancient ecosystems at deep-sea hydrothermal vents before the advent of oxygen-producing organisms. Other populations of microbes, however, might also have been present, adding to ecosystem diversity. For example, if sulfate was found in seawater, even in small concentrations (we'll hear much more about sulfate in later chapters), a process known as sulfate reduction could have been supported. In this process, sulfate-reducing bacteria gain energy and grow through the reaction of sulfate with organic matter or H_2; sulfide and CO_2 are then produced.[8] These ancient sulfate reducers, then, could have lived off the organic matter from other dead microbes, or from H_2 emanating from the hydrothermal vents. It seems probable, therefore, that ancient hydrothermal ecosystems housed quite diverse populations of microbes, but H_2 would have been the most likely primary source of energy.[9]

Deep-sea hydrothermal systems would not, however, have been the only place to encounter life on early Earth before oxygen. Indeed, the early biosphere likely also saw considerable action on land and in the upper reaches of the global ocean. There are many reasons for saying this, but we start with the H_2S (hydrogen sulfide) that was so uninteresting for life in the deep, dark, oxygen-free ocean. On land, and in the light of day, H_2S suddenly becomes very interesting indeed. When available, this H_2S would have been of great use to a group of photosynthetic organisms that evolved well before the oxygen producers we know best. Since these photosynthesizers don't produce oxygen, they are known as anoxygenic phototrophs.[10] I will say a good deal about their evolution in the next chapter, but here I will introduce you to their ecology. In fact, they are not rare.

I have always dreamed of living by the sea, and in an act of reckless indulgence, my family agreed to purchase a small house on the Danish island of Bornholm, where we vacation every summer. On the southeastern part of the island there are a series of beautiful white beaches whose sand is so fine and regular that it was used to make hour glasses. Because the sand is so fine, it dampens easily and stays wet for a long time, and if you scrape your foot along a wet patch, you will typically find a very interesting and repeatable layering of green, red, and black (plate 2). If you stick your nose down close, you sense the faint smell of hydrogen sulfide. The sulfide comes from sulfate reduction, a process we met earlier, and the black color is produced by the reaction between sulfide and small amounts of iron minerals in the sand. The green band is colored by the oxygen-producing cyanobacteria we will explore in detail in subsequent chapters. Important to our discussion here, the red band is colored by anoxygenic photosynthetic organisms. These organisms use the Sun's energy to convert sulfide to sulfate, and in the process, they generate cell biomass from CO_2.

While the Bornholm sand provides a good environment for sulfide-using anoxygenic phototrophic populations, these sands provide a rather poor analogue to the ancient Earth. This is because the sulfide used by this population of anoxygenic phototrophs is obtained, ultimately, from the decomposition of organic material produced by the cyanobacteria populating the upper layers of the sand. Take away the cyanobacteria, and there would be no sulfide for the anoxygenic phototrophs to use. A more analogous environment to early Earth would be the type of thermal springs found at Yellowstone National Park, on Iceland, or on North Island of New Zealand.

I visited Yellowstone National Park as a kid. I marveled at the hot springs and I was mesmerized by "Old Faithful" but mostly, I was absorbed by the hunt for bears from the safety of our car. I do remember the colors though, the beautiful browns, oranges, reds, and greens spreading like abstract paintings from the seemingly bottomless throats of the hydrothermal springs. Only years later did I learn that these colors were formed from mats of bacteria, many of which contain large populations of anoxygenic phototrophs that oxidize the hydrothermal sulfide emanating from the springs. This, I think, is a decent analogue to what we might have found on ancient Earth. As with deep-sea hydrothermal vents, we can imagine complex ecosystems developing. The sulfide-oxidizing anoxygenic phototrophs would produce sulfate, and this would be used by sulfate-reducing bacteria to oxidize the organic matter produced by the phototrophs. During sulfate reduction, sulfate is reduced to sulfide, recycling sulfide for use again by the anoxygenic phototrophs. As in the hydrothermal vent ecosystems, various fermenting

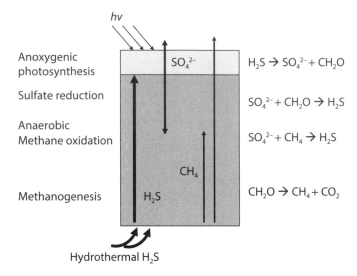

Figure 2.1. Possible workings of an ancient sulfur-based microbial ecosystem, and the likely additional influence of microbial methane cycling. Methane cycling will occur if some of the sulfate produced during anoxygenic photosynthesis is lost from the system, as through river runoff, for example. Also shown is the process of anaerobic methane oxidation, the oxidation of methane with sulfate; this is of possible significance, but not discussed in the text. Redrawn from Canfield et al. (2006). CH_2O indicates organic compounds.

bacteria would aid in organic matter decomposition and help generate food for the sulfate-reducing bacteria. This type of ecosystem is known as a "sulphuretum," a term first introduced by Laurens Baas Becking in 1925 (Baas Becking would later initiate the field of geobiology in his 1934 book, *Geobiologie*) (fig. 2.1). Such an ecosystem also cycles matter in a way directly analogous to what happens in our modern oxygenated biosphere. Just replace sulfide with water and sulfate with oxygen.

If some of the sulfate produced by the anoxygenic phototrophs washed away with the flowing hydrothermal water, then there would be insufficient sulfate to decompose all of the dead biomass via sulfate-reducing bacteria. This deficiency would allow a community of methane producers to develop and decompose the rest of the organic material, adding even more complexity to our sulfide-fueled terrestrial ecosystem. If one could have visited these ancient ecosystems, one would have marveled at the microbial stringers and colorful mats, somewhat analogous to those found today. Life would have appeared bountiful in these hydrothermal areas. However, with a general paucity of such areas on a global scale, sulfide-based ecosystems would have probably contributed only a little to the total activity of the biosphere.

To find big players on Earth before oxygen, we look to the skies and recognize that ancient volcanoes would have spewed gases like H_2, SO_2 (sulfur dioxide), CO_2, and H_2S into the atmosphere. As in the deep-sea hydrothermal systems, methane-producing populations would have been supported by the volcanic H_2 and CO_2. One can imagine methanogens living in saturated soils on land, in lakes, and in the sea, combining CO_2 and hydrogen gas. More importantly, though, many anoxygenic phototrophic bacteria can convert hydrogen gas (H_2) to water. They couple this reaction to the generation of cell biomass from CO_2.[11] These ancient hydrogen-utilizing phototrophs would have populated ancient lakes, ponds, and the surface ocean; indeed, they would have lived in any watery environment lit by the Sun where the H_2 and CO_2 from the atmosphere could dissolve. How productive might such an ancient metabolism have been? Minik Rosing, Christian Bjerrum, and I built on a model originally conceived by Jim Kasting and his group at Penn State (we met Jim in the last chapter), and we estimated that anoxygenic phototrophs using hydrogen gas could have produced biomass at a maximum rate of about 3×10^{13} moles of organic carbon per year (equivalent to 3.6×10^{14} grams of carbon per year). This sounds like a big number, but it is still one hundred times less than the rates of organic carbon production in the present biosphere, which is supported by oxygenic photosynthesis.

To find what may have been the biggest player in the ancient pre-oxygen biosphere, we take an imaginary dive with *Alvin* into the depths of the ancient oceans. As we descend, we focus our attention at that depth just where the Sun's light fades into darkness. Here, we might well observe a dense population of bacteria, and curiously, an abundance of iron oxide minerals, which you can view as basically rust. Our chemical sensors would also detect an accumulation of dissolved ferrous iron (Fe^{2+}) in deep waters below where the rust particles were found. What's going on here?

We'll hear much more about iron in the chapters to follow, but basically, ferrous iron is the form of iron that persists in the absence of oxygen, and it readily dissolves in water. Therefore, without oxygen in the atmosphere, the ferrous iron coming from hydrothermal vents would have accumulated in the ancient deep sea. Indeed, we have geological evidence for this in the massive accumulation of a peculiar type of sedimentary rock known as banded iron formations (BIFs) (plate 3). These types of rocks are particularly abundant in the rock record before about 2.5 billion years ago (we'll hear much more about BIFs in later chapters), and our best models argue that these BIFs formed from ferrous iron dissolved in seawater. But focusing again on the particle layer, what is the source of all these bacteria, and the rust?

The microbiologist Fritz Widdel from the Max Planck Institute for Marine Microbiology in Bremen, Germany, is the picture of patience. Indeed, this patience has allowed him to make many fundamental breakthroughs in our understanding of microbial metabolism. In one of these breakthroughs, Fritz and his students collected mud from a drainage ditch near the Max Planck Institute,[12] and incubated it in the light with added ferrous iron. They waited and waited and waited, and eventually, after some months, they saw a population of purplish bacteria growing on the mud. They transferred these bacteria to new media, and waited some more. Then finally, after the waiting was done, they had isolated a population of anoxygenic phototrophic bacteria (plate 4) able to grow from the ferrous iron and forming, essentially, rust in the process. People had long suspected that such a population might exist in nature, but no one had either Fritz's patience or talent to isolate them. But, would such a population have been of importance in the ancient oceans?

As students of the distant past, we try to identify modern environments that resemble, as closely as possible, the ancient environments we seek to understand. Where, though, might we find an analog to a three to four billion-year-old iron-rich ocean? While I was pondering this, a PhD student from Canada, Sean Crowe, was visiting our lab and explained that he was working on Lake Matano, Sulawesi, Indonesia. This turns out to be just such an environment. The lake is deep, almost 600 m, clear and stable. Critically, it accumulates substantial concentrations of ferrous iron in the bottom waters. Sean was planning another trip to Matano and invited my PhD student CarriAyne Jones to join. Sean and CarriAyne learned much about the lake, but critically, they discovered an anoxygenic phototrophic population just where iron,

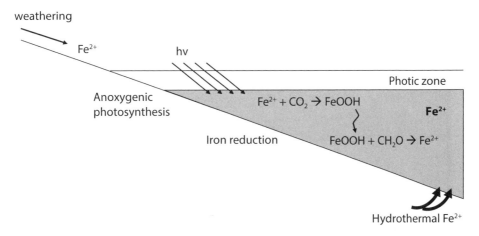

Figure 2.2. Possible structure of an Fe-based ecosystem in the oceans. See text for details. Redrawn from Canfield et al. (2006).

coming from below, is oxidized as it meets the fading light from above. While these phototrophs have evaded isolation, Sean and CarriAyne concluded, after a variety of considerations, that they most likely oxidize ferrous iron to rust in the lake. Lake Matano is a long way from Fritz Widdel's ditch, but both environments point to the possibility that iron-oxidizing anoxygenic phototrophic bacteria could have been important contributors to the biological productivity of early Earth.

We imagine that these iron-oxidizing organisms would have been partners in ecosystems involving fermenting bacteria and also so-called iron-reducing bacteria. Iron reducers are a well-known group of microbes who grow by reducing iron oxide back to ferrous iron, and they oxidize organic matter or H_2 in the process. They would have done the same job in the ancient oceans by recombining the products of photosynthesis, the iron oxides, and the cell biomass, oxidizing the cell biomass back to CO_2, and reducing the iron oxides back to dissolved ferrous iron (fig. 2.2). This ecosystem would effectively recycle the iron, and the activity of the phototrophs would ultimately be controlled by the availability of ferrous iron and nutrients. Minik, Christian, and I also attempted to model the activity level of this ancient phototrophic population. Our calculations were highly imprecise and fraught with many uncertainties and assumptions,[13] but recognizing these issues, we calculated that an iron-based biosphere could have been, perhaps, 10% as active as today's marine biosphere. Not too shabby.

Let's now summarize. It seems likely that before the evolution of oxygen-producing organisms, numerous interesting and dynamic ecosystems existed on Earth in environments ranging from the deep sea, to the surface ocean, and to land-based hydrothermal systems, as well as in lakes and in soils. In some instances, these ecosystems would have been rather diminutive, and in other instances, quite obvious to the naked eye. The most active ecosystem may have been based on the oxidation of dissolved ferrous iron in the oceans. It seems most likely, however, that the ancient biosphere would have been much less active than the present one, which is driven by oxygen-producing organisms.

Can we find any evidence of such ancient ecosystems in the geologic record? This is far from certain. One difficulty, as we shall explore in more detail in chapter 7, is that we're really not quite sure when oxygen-producing cyanobacteria first evolved. Therefore, we are not certain how far back in time we must go to be confident that the rocks we explore were deposited in a world free from oxygen-producing organisms. And even so, there aren't many of these ancient rocks around, and those that we can find are not in good shape. As we will explore in chapter 6, the same plate tectonics that make the world such a habitable place, leave a fragmented and compromised geologic record. The tossing, turning, and rumblings of the planet promote the erosion and weathering of ancient rocks and also encourage their burial, heating, and deformation. In short, most of the rocks once found at Earth's surface are gone, and many of those that remain have been heated and badly deformed, and this problem grows worse the further back in time we look.

But still, despite the insults of age, there are some clues to the types of microbial life that occupied very ancient ecosystems. Indeed, a most revealing place to look is the nearly 3.5-billion-year-old rocks from "North Pole," Australia.[14] These rocks are in remarkably good shape for their age, and they have been studied intensively by my Australian colleague Roger Buick, now of the University of Washington. Roger paints a picture of an active volcanic terrain at the ocean's edge. Imagine semi-isolated marine lagoons washing over basaltic outpourings, and accumulating sediment in an environment with locally high concentrations of sulfate in the water. The sulfate was likely sourced, ultimately, from sulfur gases discharging from the local volcanoes. Sulfate minerals, at least some of which were originally precipitated as gypsum (they are now all barite, $BaSO_4$), are associated with fine-grained pyrites and form an important component of the sediments. My then postdoc Yanan Shen, together with Roger Buick and I looked at the sulfur isotope composition of the sulfate and the pyrites in these rocks. I will say more about what these isotopes tell us in chapters 7 and 9, but our results suggested that the sulfide in the pyrites formed from microbial sulfate reduction. We were quite proud of this result because it documented both the early evolution of sulfate-reducing bacteria and the earliest specific microbial metabolism in the geologic record.

As one might expect, this finding came under intense scrutiny, with the discussion centering on whether the sulfide was instead formed from a thermochemical process not involving organisms at all. This can happen if you heat sulfate up with organic matter to a sufficiently high temperature. Yanan Shen has since revisited these rocks with a more sophisticated sulfur isotope approach and has generated new results that are consistent with our earlier finding.

There's still more. The basaltic rocks lying just below the sediments explored by Yanan, Roger, and me are crisscrossed with a number of silica-rich dikes, and these contain tiny inclusions of fluid and gases. Since the dikes formed at about the same time as the sediments, these inclusions could, in principle, hold further clues as to the microbes living some 3.5 billion years ago. Indeed, Yuichiro Ueno from the Tokyo Institute of Technology, along with his colleagues, looked at the gases present in these inclusions and found that many contained a good deal of methane. They divided the inclusions into those that looked primary, dating back to the time that the dikes were formed, and those that looked secondary, having formed after the original emplacement of the dikes. They then measured the isotopic composition of the methane, as biogenic methane yields a distinct isotopic signal. They found that the primary inclusions contained a methane isotopic composition (don't worry about the details) consistent with a biogenic origin through microbial methanogenesis. In contrast, the secondary inclusions had methane isotopic compositions more consistent with nonbiogenic sources. This is a beautiful piece of geochemical sleuthing and shows that methane-producing organisms were also part of the ancient microbial ecosystem at North Pole, Australia.

All in all, geological evidence suggests that many of the processes that we have imagined were part of the early biosphere that was in place 3.5 billion years ago. These processes include methanogenesis, sulfate reduction, and decomposition of dead organic biomass, which was likely aided by a host of different fermenting bacteria. Unfortunately, indicators of anoxygenic photosynthesis in rocks so old are not terribly robust, so the geologic record is rather silent on the antiquity of this process. We will, however, explore other ways of looking at the antiquity of anoxygenic photosynthesis in the next chapter.

There may, however, be another way to investigate the early history of microbial evolution independently of the geologic record. The premise is simple. All organisms on Earth contain a record of their evolutionary history in their DNA. This is because the DNA of any organism, including us, is the product of all of the changes that have occurred in its lineage before the present time. The history recorded in DNA, however, is complicated. It is influenced by a number of factors, including the duplication of genes, the birth of new genes, the loss of genes, the transfer of genes between organisms, and all the mutations accumulated in the DNA sequences through time. In principle though, we can compare the DNA of one organism to that of another to understand how the DNA differs between them, and if we compare enough different organisms, we can understand the history of DNA evolution. Eric Alm from the Massachusetts Institute of Technology (MIT) and his student Lawrence David have taken a particularly sophisticated look at this issue. They analyzed nearly 4000 genes from over 100 organisms and by doing so, they offer a history of gene evolution. The results are spellbinding.

I have reproduced some of these results in plate 5. What we have is a history of the importance of different metabolic types through Earth's history. There are many assumptions behind the construction of such an evolutionary history, and this is also among the first attempts at this fascinating approach. With all of this in mind, I stare at the results. This data is, in principle, exactly what we are after. What type of organisms defined the biosphere 3.5 billion years ago? When did sulfate reduction evolve? Methanogenesis? The results in plate 5 suggest that the sulfur cycle was quite active early in Earth history. Autotrophic metabolisms also evolved early, and oxygen-utilizing genes expanded more recently. According to this analysis, however, methanogenesis is a metabolism that evolved later. Oops. I find this hard to believe, especially given the geological evidence described above, but as I said, we are looking at the initial application of this approach, not the last.

Let's try to pull all of this together. By all reckoning, Earth enjoyed an active and diverse biosphere well before the evolution of oxygen-producing cyanobacteria. This biosphere was fueled, mainly, by chemical compounds liberated during volcanism, underscoring again the importance of plate tectonics in shaping life on our planet. The geologic record provides support for an early diverse biosphere, as do new molecular approaches aimed at deciphering evolutionary history from DNA sequences. It seems likely, though, that this early biosphere was much less active than what we enjoy at present. In the next chapters we will begin to look at the steps leading to the evolution of oxygen-producing organisms and will begin to understand how the present biosphere was formed.

CHAPTER 3
Evolution of Oxygenic Photosynthesis

The race was on, but nobody really knew there was race, at least not at first. The year was 1771 (or maybe 1772), and the Swedish pharmacist Carl Wilhelm Scheele was very busy indeed. Having just begun as laboratory assistant to the chemist Torbern Bergman of Uppsala, Scheele was immersed in unraveling the mysteries of air. His real motivation, however, was more boyish than this; he really wanted to understand the nature of fire. Scheele admitted, "one could not form any true judgment regarding the phenomena fire presents, without a knowledge of the air."[1]

At this time in the history of science, air was a truly remarkable and mysterious substance. We can see this by hearing how Scheele himself summarized its properties: "Air is that fluid invisible substance which we continually breathe, which surrounds the whole surface of Earth, is very elastic, and possesses weight." In addition to these meager facts, it had been demonstrated that air contained carbon dioxide, but other than this, precious little was known. Part of the problem was that the concept of chemical elements was just emerging, and those making up the components of air had yet to be discovered. Another problem was that the nature of air could not be decoupled from the popular (at the time) phlogiston theory. Phlogiston was believed to be a colorless substance without mass, which was liberated by flammable substances as they burned. Therefore, air became "phlogistated" when objects burned, and flammable substances could no longer burn when sufficient phlogiston accumulated in air. This explanation, however, was likely insufficient to Scheele, and he realized that the real key to understanding the burning processes was to understand the nature of air.

One can follow Scheele's thought process in his book *Chemical Treatise on Air and Fire,* published in 1777 and now translated into English. In early experiments, Scheele burned, among other things, a variety of sulfurous compounds to reaffirm one property of air that was already known. That is: "Substances which are subjected to putrefaction or to destruction by means of fire diminish, and at the same time consume, a part of the air." From this Scheele concluded that "air must be composed of elastic fluids of two kinds." But what was the nature of these two kinds of fluids? The experiments continued. The real breakthrough came when Scheele distilled potassium nitrate (saltpeter) in a mixture of nitric and sulfuric acids (nasty stuff indeed!) and collected gases near the end of distillation, when "blood red vapours are produced."[2] Lo and behold, if a candle is introduced into this gas, "not only will it continue to burn, but this will take place with a much brighter light than in ordinary air." He produced the same product in a variety of other ways and called this product "fire-air." He then estimated that it makes up about one-third of our atmosphere.[3] What Scheele found, of course, was oxygen. Through a variety of further ingenious experiments with plants, rats, and bugs, he went on to surmise that "fire-air" is replaced by (converted to in his view) carbon dioxide through respiration, and that "fire-air" is absorbed by the lungs of animals and is transported by blood through the body.

Scheele discovered that air consists of oxygen, carbon dioxide, and a major portion of an unreactive substance; he called this "vitiated air," which we now know to be nitrogen gas. As for "fire-air," Scheele mused, "I am inclined to believe that fire-air consists of a subtle acid substance united with phlogiston." This conclusion, however, must have seemed wanting because in September 1774 Scheele sent a letter to the world-renowned French chemist Antoine Lavoisier explaining his experiments and asking for advice.

Ironically, at the same time and unknown to Scheele, the Englishman Joseph Priestley was in Paris discussing his own experiments on oxygen production face to face with Lavoisier.[4] Lavoisier apparently listened closely, because he quickly produced oxygen on his own; not bound to the trappings of the phlogiston theory, he correctly recognized oxygen as an element and gave it the name we now know (meaning "acid producer"). He too, like Scheele, explored the role of oxygen in respiration, and more accurately than Scheele, he estimated that oxygen comprises about 25% of Earth's atmosphere.

Many have complained that Lavoisier gave scant credit to Priestley and no acknowledgement to Scheele in his own writings about oxygen. Indeed, Lavoisier claimed never to have received Scheele's letter, although it surfaced among Lavoisier's wife's effects in the 1890s. Did Lavoisier receive the letter but bury it in his wife's things so he could claim credit for the discovery of oxygen? Or, did his wife receive the letter first and hide it so that Lavoisier could claim credit without the knowledge of a competitor?[5] We will never know, but clearly, as often happens in science, our ability to explain the world within an existing paradigm becomes so inadequate that new thinking is necessary. Clearly, the time was ripe in the early 1770s for the discovery of oxygen.

Let's continue with our brief historical digression. Although Scheele experimented extensively with plants, he never uncovered the process of photosynthesis. Priestley did, however, and in 1771, even before he discovered oxygen, he had found that plants provided a substance that could sustain a mouse and the burning of a candle. Most, however, credit the Dutch physician Jan Ingenhousz for fully understanding how photosynthesis works. By 1779, Ingenhousz was able to summarize his discovery, in surprisingly modern terms: "It will, perhaps, appear probable, that one of the great laboratories of nature for cleansing and purifying the air of our atmosphere is placed in the substance of the leaves, and put in action by the influence of the light."[6] We can see that by the end of the 1700s, there was a reasonably good understanding of the major chemical composition of the atmosphere, including the origin of oxygen and its role in respiration.

In the remainder of the chapter, we will continue to explore history, but of another kind. Our goal will be to understand the evolution of photosynthetic oxygen production on Earth. I hope we can agree that this event was one of the major transforming events in the history of life. With no oxygenic photosynthesis, there would be no oxygen in the atmosphere; there would also be no plants, no animals, and nobody to tell this story.

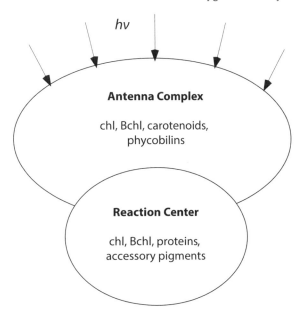

Figure 3.1. Light harvesting in a photosynthetic organism. Figure shows the relationship between the antenna complex and the reaction center(s), each containing light-collecting pigments. Figure redrawn from Canfield et al. (2005).

As a biological innovation, where did oxygen production come from? To unravel this problem, we must begin by looking at how oxygen production by photosynthesis works. Everything starts with light, and in figure 3.1, I have reproduced the salient features of how light (*hv*) is captured during the photosynthetic process. This general picture pertains to all types of photosynthesis, including photosynthesis by cyanobacteria, the first oxygen producers (as we explore in chapter 4). It also pertains to plants, or even the anoxygenic phototrophs we met in the last chapter. To collect light, all phototrophs use an "antenna complex." This is made up of a variety of pigments that capture the light (fig. 3.1). Once captured, the light's energy is transferred to the reaction centers. These constitute the business end of the process where, in the case of oxygenic phototrophs, oxygen is made. Nature has made things complicated, and in oxygen producers, there are actually two coupled reaction centers: one is known as photosystem I (PSI) and the other as photosystem II (PSII). They are coupled together in what is commonly called the Z-scheme, and as I hope becomes obvious below, this arrangement makes a lot of sense.

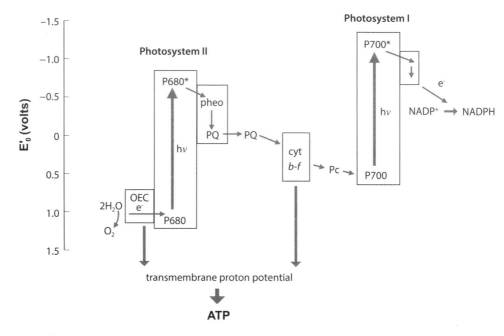

Figure 3.2. The coupled reaction centers in oxygenic photosynthesis, the so-called Z-scheme. The trick in understanding this is to follow the electrons as explained in the text. Figure kindly provided by Raymond Cox.

To understand how oxygenic photosynthesis works, our simple goal will be to follow electrons. The process is outlined in figure 3.2, and we start with PSII. Energy from the antenna complex is transferred to a special chlorophyll molecule called P680. This energy causes P680 to develop an excited state, named P680*, which becomes a strong reductant (source of electrons). Like a child screaming for the bathroom, P680* easily dispenses with an electron and hands it over to a chemical compound known as a pheophytin, reducing the pheophytin in the process. We will worry about what happens with the pheophytin in a minute, but our immediate concern is the P680 chlorophyll that has just lost an electron. This electron must be replaced or the whole process will run down and stop.

In the remarkable innovation of oxygenic photosynthesis, the electron comes from water. This is a big deal because, as we know from experience, water is a stable substance; we don't worry about it bubbling off oxygen, for example, when we take a bath. The beauty of P680 is that after it loses its electron, it becomes a very strong oxidant (an electron acceptor, the strongest known in nature in fact), which means that it can take electrons from water, producing oxygen as a byproduct. This all sounds easier than it really is. Getting the electrons to flow from water to P680, with the formation of oxygen, requires a carefully orchestrated biochemical ballet. This dance is conducted by the so-called oxygen-evolving complex (OEC), which at its core contains a four-membered Mn (manganese) cluster. We will hear more about this shortly.

But first, back to the pheophytin. When it obtains its electron from P680*, it does so quickly (within 3 picoseconds in fact; that's 3 trillionths of second!); this is critical or the electron would recombine again with the oxidized P680, which, as we saw above, is a strong oxidant and is very good at extracting electrons. If it did so, the whole process would be cut short and stop. Ultimately, the cell needs to put the electron in the pheophytin into a soluble electron carrier known as NADP(H).[7] Packed in NADP(H), the electron can be used to conduct all kinds of biochemistry for the cell. However, the pheophytin is barely able (barely reducing enough in chemical terms) to put its electron into NADP(H), and if it did so now, the cell would gain very little from the photosynthetic process.

Instead, the electron is used to do work. In what happens next, we can imagine the electron sliding downhill, like a go-cart on a fine summer day. It rides from the pheophytin to a quinone molecule and further downhill to a series of other proteins. The cell takes advantage of this carefree ride, and as the electron rolls downhill, ATP is formed.[8] ATP is the energy currency of the cell, and in desperately unromantic terms, one could make the case that the purpose of life is to make ATP. Anyway, after the electron makes it to the bottom of the hill, there isn't much energy left in the protein holding it. There is no way it can make NADP(H) and no way it can do much else for the cell. So, the electron is handed over to PSI (photosystem I). Here, there is another chlorophyll molecule known as P700, whose oxidizing form lies in wait for this electron. As P700 and the electron unite, energy from the antenna complex (*hv*) propels P700 to an excited and reduced state, P700*, which is, in fact, a much stronger reductant than the excited P680* we just discussed. The go-cart ride begins again, but because the electron starts from a higher level (a more reduced state), it remains sufficiently reducing (electron yielding) to easily combine with NADP$^+$ to form NADP(H).

We are not quite finished. If we just stop here, the electrons packed in NADP(H) would simply accumulate. This cannot and does not happen. Indeed, the electrons in NADP(H), which, if you remember, were ultimately sourced from water, are transferred to CO_2, producing the organic compounds used to build cells. This process is known as carbon fixation, and the incorporation of CO_2 is promoted by an enzyme known as Rubisco.[9] Indeed, Rubisco-promoted carbon fixation is the basis for virtually all the food we eat and nearly all of the fossil energy we use. It's been a long ride indeed, but in the end, the cell gets just what it needs to grow, and it spits out O_2 as a waste product in the process.

With this brief description as a backdrop, we will approach the evolution of oxygen-producing organisms by considering the evolution, and assembly, of its basic constituent parts. We will focus on the following key questions: (1) What is the evolutionary history of chlorophyll? (2) What are the evolutionary histories of PSI and PSII? (3) What is the origin of the oxygen-evolving complex in PSII? And finally, (4) what is the evolutionary history of Rubisco? In addressing these, we will seek to understand the complex path leading to the evolution of oxygenic photosynthesis on Earth.

As hinted at in the last chapter, oxygenic photosynthesis was probably not the first type of photosynthesis; that honor goes to anoxygenic photosynthetic bacteria. I will not use much time exploring the evolution of those earliest photosynthetic organisms, and indeed, not much is known. However, as we shall see below, the evolution of oxygenic photosynthesis makes most sense when viewed in the light of anoxygenic phototrophs as precursors. Thus, we will look to anoxygenic phototrophs more than once to find clues to the origin of oxygenic photosynthesis.

We begin with chlorophyll. I remember well my mom's frustration as I tried to sneak through the back door with bright green stains on my new jeans or, even worse, on my white Sunday shirt. Other than challenging mothers (and fathers) to keep their kid's clothes in a reasonable state of shine, green-staining chlorophyll, as we saw above, also serves several different functions in an oxygen-producing organism; it is an integral part of the antenna complex and it is a critical component of both photosystems. The main importance of chlorophyll for oxygenic photosynthesis, however, is that it produces the highly oxidizing form of P680 in PSII, which can extract electrons from water. This is the main innovation of oxygenic photosynthesis.

So, where does chlorophyll come from? As it turns out, chlorophyll is not a particularly weird molecule at all. It is related structurally and chemically to a variety of other very common molecules (so-called porphyrin molecules) widely used in all sorts of cellular enzymes, including the heme in hemoglobin. Chlorophyll is also closely related to the bacteriochlorophylls used in anoxygenic photosynthetic organisms. Indeed, the synthesis pathways of chlorophyll *a* and the various bacteriochlorophylls are very similar and diverge mainly in the very last steps. Therefore, the biochemical distance is small between chlorophyll, bacteriochlorophyll, and even the common porphyrins used in our cells. The question, though, is which came first.

Most biochemists would agree that porphyrins, as a general class of molecules, evolved before chlorophyll and bacteriochlorophyll. These early porphyrin molecules would have helped promote the biochemistry of the earliest life on Earth. If we now consider the photosynthetic pigments chlorophyll and bacteriochlorophyll, we might guess that chlorophyll evolved first—that is, if we base this assessment on the formation pathways of these molecules. This idea was presented by Sam Granick in 1965 (known as the Granick hypothesis) and was based on the idea that chlorophyll *a* forms in only one step from its immediate precursor, a molecule called chlorophyllide *a*, whereas several steps are needed to form bacteriochlorophyll *a* from the exact same precursor molecule. Thus, chlorophyll *a* is easier to make.

We can look at this problem, however, from another angle. In this era of genomics, evolutionary histories can be constructed directly from the sequences of DNA within organisms. We saw an elegant application of this at the end of the last chapter, where David and Alm explored the evolutionary history of genes that conduct a wide variety of different microbial metabolisms. Using a related approach, Jin Xiong and Carl Bauer from Indiana University explored the evolutionary history of oxygenic and anoxygenic photosynthetic organisms. As noted above, chlorophyll and bacteriochlorophyll are formed by pretty much the same biochemical pathway up to the very last stages of biosynthesis. Therefore, genes can be identified from both oxygenic and anoxygenic phototrophic organisms that conduct exactly the same process in the synthesis of both chlorophyll and bacteriochlorophyll molecules. Xiong and Bauer determined the DNA sequences of several of these genes, and when comparing sequences from among the same genes, those from anoxygenic phototrophic bacteria always appeared more ancient than those from oxygenic phototrophs. This is good evidence that chlorophyll biosynthesis is more modern than bacteriochlorophyll biosynthesis, suggesting that bacteriochlorophyll came first. Therefore, the fact that chlorophyll has a somewhat simpler final biosynthetic pathway must be balanced against other evidence that bacteriochlorophyll, in fact, originated first.

Thus, our idea that anoxygenic photosynthesis predates oxygenic photosynthesis seems pretty well supported, as does the idea that chlorophyll biosynthesis emerged later, perhaps together with the evolution of oxygenic photosynthesis. But why? Why did organisms first use bacteriochlorophylls, the more complex biosynthetic product, rather than starting off immediately with chlorophyll? It could be by chance that the bacteriochlorophyll pathway evolved first, and since it worked, there was no evolutionary pressure to incorporate a new pigment system. Or it could be, as Martin Hohmann-Marriott from the University of Otago and Bob Blankenship from Washington University (more on Bob below) pointed out, that the early enzymes forming bacteriochlorophyll could have conducted multiple steps in the synthesis process, essentially bypassing the chlorophyll step, or producing chlorophyll as only a minor product. Only later, as the genes evolved, did chlorophyll synthesis become a major formation pathway in oxygen-producing organisms. This seems possible, but the reality is that we still have much to learn.

We now turn our attention to the reaction centers, PSI and PSII, the heart of the oxygenic photosynthetic process. Bob Blankenship of Washington University has studied photosynthesis for many years and he, together with his former student Jason Raymond, have probably thought about the evolution of photosynthesis as much as anyone. As they say, pictures tell a thousand words, and long ago Bob found similarities in function and structure between PSI and PSII, and the reaction centers used by anoxygenic phototrophs. Let's try to see what Bob saw, by reference to figure 3.3. Bob compared the biochemical pathways of PSI and PSII to those of anoxygenic phototrophs. Through this com-

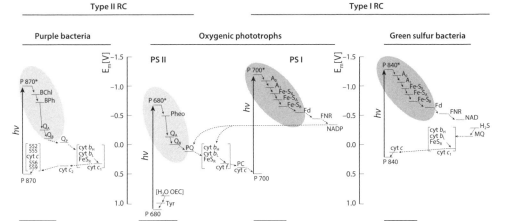

Figure 3.3. Comparison of the reaction centers in anoxygenic phototrophs with those in the coupled photosystems of oxygen producers. Don't fret about the details, but note the similarities between PSII and the Type I reaction center of the purple bacteria, and PSI and the Type II reaction center of the green sulfur bacteria. Borrowed with slight modification from Blankenship (2010). Reproduced with permission.

parison, Bob could see great similarities between PSI and the "Type 1" (or FeS type) reaction center used by some anoxygenic phototrophs, including a group known as "green sulfur bacteria" or affectionately known as GSBs.[10] Critically, in both PSI and the Type 1 reaction center, the electrons are shuttled through a similar series of proteins after they emerge from the highly activated chlorophyll or bacteriochlorophyll molecules we described earlier. When Bob looked at PSII, he also saw many similarities with the "Type 2" (or Q type) reaction centers used by other anoxygenic phototrophs. This Type 2 reaction center has structural similarities to PSII and also shuttles electrons in a similar manner. Many of the anoxygenic phototrophic organisms using the Type 2 reaction center are found in a group of *Bacteria* known as purple bacteria.[11] These purple bacterial anoxygenic phototrophs are a very diverse group, but they all use the same type of reaction center.

As convincing as this might be, Bob and Jason, together with Sumedha Sadekar, have taken the comparison of the reaction centers to another level. In the last several years, technology has allowed the determination of large protein structures to uncanny resolution, so that in many cases, one can make out individual atoms. With this resolution, the similarities

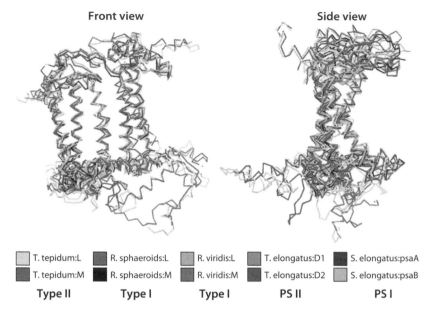

Front view　　　　　　　　　　**Side view**

☐ T. tepidum:L	■ R. sphaeroids:L	☐ R. viridis:L	■ T. elongatus:D1	■ S. elongatus:psaA
■ T. tepidum:M	■ R. sphaeroids:M	■ R. viridis:M	■ T. elongatus:D2	☐ S. elongatus:psaB
Type II	**Type I**	**Type I**	**PS II**	**PS I**

Figure 3.4. Comparison of the structures of the photosynthetic reaction center proteins. Also indicated is whether the proteins are Type I or Type II anoxygenic photosynthetic reaction centers, or whether they are PSII or PSI from oxygenic phototrophic cyanobacteria. Borrowed with slight modification from Sadekar et al. (2006). Reproduced with permission. See plate 6 for a color version.

and differences in structure between the different reaction center proteins can be compared with certainty. The logic in making such comparisons is that proteins with similar structures are more closely related evolutionarily. So, when structures of the proteins from PSI, PSII, and Type 1 and Type 2 reaction centers are compared (fig. 3.4; see plate 6 for a color version), there is a remarkable similarity, particularly in the middle regions of the protein where it is embedded in a membrane within the cell. This is good evidence that all of the reaction center proteins are related to one another. With further analysis of these protein structures, Sadekar, Raymond, and Blankenship concluded, consistent with Bob Blankenship's original suggestion, that anoxygenic photosynthetic Type 1 and Type 2 proteins are ancestral to PSI and PSII.

You can probably guess where this is going. A logical explanation for all of these observations is that the reaction centers PSI and PSII were derived from the Type 1 and Type 2 reaction centers already existing among anoxygenic photosynthetic bacteria. Somehow, and the details are not at all clear, these two pre-existing reaction centers were connected to form the coupled photosystems in oxygenic photosynthetic bacteria. The connection may have occurred as the reaction center of one type was transferred into an organism containing the reaction center of the other type; alternatively, the two types of reaction centers were coupled already in an original anoxygenic photosynthetic organism, which later developed phototrophic oxygen production. In either of these scenarios, anoxygenic phototrophs once contained both types of photosystems. If the first scenario is true, then the precursor anoxygenic phototrophic organism, with the two photosystems, has apparently been lost from nature (but, see below). If the second scenario is true, then modern GSBs and purple bacteria have each lost one of the two reaction centers. Bob Blankenship argues that it's impossible at present to distinguish between the two possibilities, and perhaps it is. However, John Allen from the University of London likes the idea that the two reaction centers once existed together in the same anoxygenic phototroph, perhaps derived from a single precursor through a gene duplication event.[12] In his view, each of the photosystems was of service to the organism, and either one or the other was chosen for use by the organism depending on environmental conditions. At some point these two photosystems combined. In Allen's view, anoxygenic phototrophs may well exist today with both reaction centers intact. It would certainly be exciting to find such an organism!

While we have come closer to understanding the evolution of oxygenic photosynthetic organisms, we have yet to explain how oxygen came to be produced. Here, we need to focus on the oxygen-evolving complex (OEC), containing in its core a cluster of 4 Mn atoms and a Ca (calcium) atom. Just how the whole thing works is still under active debate, but it is well known that Mn atoms are the workhorses here. Basically, these atoms act as a biological capacitor. In order to form O_2, the oxidized form of P680 removes 4 electrons from the Mn-complex (not directly from water). These 4 electrons are replaced as two water molecules liberate electrons, and thereby form O_2. Curiously, the oxygen-evolving complex, and in particular the manganese cluster, is unique to oxygenic phototrophs. Nothing like it is known anywhere else in biology. Jason Raymond and Bob Blankenship, however, have some ideas as to where it might have come from. They have developed new methods, hinted at above, for comparing the structures of different proteins and have hypothesized that the four-membered manganese cluster may have derived from two-membered manganese clusters found in some other classes of proteins. Indeed, when they compared the oxygen-evolving complex with the protein Mn catalase, the similarities in structure were found to be reasonably good.

Manganese catalase is a protein used to convert hydrogen peroxide, H_2O_2, into water and oxygen. It is a detoxifying protein, since hydrogen peroxide is a harmful compound to organisms. As first suggested by Bob Blankenship and Hyman Hartman, and picked up again by Raymond and Blankenship, the manganese cluster from Mn catalase was incorporated into an anoxygenic phototrophic bacterium sometime after the evolution of chlorophyll synthesis. This assembly would have generated the first oxygen-producing organism. This organism, however, would have conducted a two electron transfer from peroxide to form O_2, rather than the more challenging four electron transfer from two water molecules to form oxygen, which we see today in plants and cyanobacteria. At some point, there was a duplication of the two-Mn cluster to form a four-Mn cluster, and this eventually allowed the evolution of oxygenic photosynthesis as we know it today.

Not everyone is happy with this hypothesis, with the main criticism being that hydrogen peroxide was probably pretty rare before the evolution of oxygen production. John Allen and William Martin have another idea and argue that the precursor oxygenic phototroph obtained its electrons from the photooxidation of manganese ions (in the chemical form Mn^{2+}) freely dissolved in ancient anoxic seas. The electrons obtained this way would have entered PSII, been transferred to PSI, and finally to CO_2. This is similar to modern oxygenic phototrophs but with the difference that the source of electrons was found outside of the cell. In this model, the Mn ions eventually moved into the cell and became held in place within PSII, forming the Mn complex. In the final stage, H_2O became the ultimate source of electrons.

We will wrap up our discussions on the evolution of oxygenic photosynthesis by looking at the final stage of the process, in which electrons are combined with CO_2 to make organic matter. This step is accomplished through a cycle of biochemical reactions known as the Calvin-Bassham-Benson cycle (or more commonly, just the Calvin cycle), where the formation of organic matter from CO_2 is promoted by the enzyme Rubisco, which we met earlier in the chapter. Indeed, Rubisco is the most abundant enzyme on the planet. While it is found among oxygen-producing cyanobacteria and plants, it is also found among a whole suite of anoxygenic photosynthetic bacteria (mainly the "purple bacteria") as well as all sorts of other bacterial species that also convert CO_2 to organic matter.

One of the fascinating things about Rubisco is that despite being the most abundant enzyme on Earth, it isn't terribly good at what it does, at least when used in oxygenic phototrophic organisms. We can identify a number of difficulties. First, Rubisco has an extremely slow turnover rate of about 0.2 to 0.3 seconds, making it one of the slowest enzymes known; this partly explains why there is so much of it around. Second, its affinity for CO_2 is rather low, although biology has provided some "fixes" for this challenge, as we shall see. Lastly, and I think the most interesting, Rubisco is in constant competition with itself. As we have discussed, Rubisco fixes CO_2 into organic compounds, and this is known as carboxylase activity (the "c" in Rubisco). It also conducts a rather stupefying oxygenase activity (the "o" in Rubisco) whereby oxygen reacts with an intermediate compound, eventually forming CO_2 again, and essentially undoing the carbon-fixation process. Thus, as if it can't make up its mind, Rubisco is characterized by competing reactions. This isn't trivial, because the oxygenase activity of Rubisco is estimated to reduce net rates of carbon fixation by 25% to 40% in plants like rice, wheat, and soybeans.

The ratio between the carboxylase activity (the favored reaction) and the oxygenase activity (the unfavored reaction) depends on the ratio of CO_2 to O_2 at the Rubisco active site. As you might imagine, the higher the CO_2 to O_2 ratio, the more favored is the carboxylase activity. The sensitivity of Rubisco to oxygen also depends on the exact type of Rubisco present. Thus, Rubiscos used by anaerobic organisms, like anoxygenic phototrophs, tend to have high oxygenase activities for the same amount of CO_2, but this is of no consequence to these organisms as they rarely see any oxygen. This observation, however, provides a possible window into understanding how oxygenic phototrophs came to use Rubisco as their carbon-fixing enzyme. The story goes something like this. Rubisco is an ancient carbon-fixing enzyme that evolved long before oxygen-producing organisms. These ancient Rubiscos had an inherent oxygenase activity;[13] this was an unintentional property of the enzyme but of no consequence since no oxygen was present. The ancient Rubisco, possibly first present in anoxygenic photosynthetic purple bacteria, was adopted by the earliest oxygenic phototroph at a time when atmospheric oxygen concentration was very low.[14] As oxygen concentration rose, both in local environments and in the atmosphere,[15] organisms developed a number of strategies to deal with the shortcomings of Rubisco's oxygenase activity. These remedies include the evolution of less oxygen-sensitive Rubiscos and the development of various strategies to concentrate CO_2 within the cell, raising the ratio of CO_2 to O_2. Such strategies are widely used by plants and cyanobacteria today.

Thus, Rubisco provides a beautiful example of how evolution does not lead to higher-order perfection. In this case we see that once Rubisco was adopted as the carbon-fixing enzyme for oxygenic phototrophs, this path was taken, and subsequent evolution through natural selection has acted to minimize the expression of Rubiscos imperfections, rather than to find a replacement. But its properties are still not ideally suited to the task.

We also see that it makes sense to view the evolution of oxygenic photosynthesis in the broader context of the evolving ancient biosphere. We can see, for example, where many of the parts comprising the oxygenic photosynthetic process were derived. Viewed in this context, it is not surprising that oxygen evolution took some time to develop and why it was not part of the original biosphere as we explored in the last chapter. Oxygenic photosynthesis was not, however, assembled as a puzzle from a number of preexisting pieces. Many unique biological innovations, like the manganese cluster and chlorophyll biosynthesis, for example, were integral parts of its evolution. Could other pathways have led to oxygenic photosynthesis? Perhaps. This, I think, is a fascinating question, and imagining alternative pathways to phototrophic oxygen production may help us visualize how this process could possibly have developed elsewhere, beyond Earth.

CHAPTER 4
Cyanobacteria: The Great Liberators

Try to imagine something so profound, so fundamental, that it changed the whole world. Think of something so revolutionary, that it forever changed the chemistry of the atmosphere, the chemistry of oceans and the nature of life itself. What about the Great Plague, the Renaissance, or World War II? These were important events indeed, and they all changed the course of human culture, but their influence outside the human realm was small. What about the extinction that killed the dinosaurs 65 million years ago, or the great Permian extinction some 250 million years ago, which laid waste to perhaps 95% of all animal species on the planet? We're getting closer. Each of these major extinctions indeed changed the course of animal evolution, but still, they did not fundamentally alter the fabric of life or the surface chemistry of Earth. What then, you might ask, did?

It was the evolution of cyanobacteria. These tiny unassuming creatures, which we met in the last chapter, completely changed everything. As we discussed, the evolution of cyanobacteria brought the biological production of oxygen to Earth for the first time. This led, in turn, to the eventual accumulation of oxygen in the atmosphere and to the widespread evolution of oxygen-utilizing organisms. These points will be explored in detail in subsequent chapters. The importance of cyanobacteria, however, goes beyond this. As we have learned, cyanobacteria were the first photosynthetic organisms on Earth to use water as a source of electrons. Unlike the sulfide, Fe^{2+}, and H_2 used by anoxygenic photo-trophic organisms, water is almost everywhere on the planet surface. This means that biological production on Earth was no longer limited by the electron source (water in this case), but rather by nutrients and other trace constituents making up the cells. In the end, the use of water in photosynthesis resulted in an increase in rates of primary production on Earth by probably somewhere between a factor of ten to a thousand, as outlined in chapter 2.[1] For the first time, life on Earth became truly plentiful. With the evolution of cyanobacteria, Earth was on its way to becoming a green planet.[2]

Cyanobacteria would have inhabited all manner of lakes, ponds, streams, and puddles on land where water persisted for at least short periods of time, as well as the whole upper sunlit layer of the oceans (typically called the photic zone). If we imagine ancient explorers of the solar system probing early Earth, they would have needed a micro-scope to find much evidence for life before the evolution of cyano-bacteria.[3]After this, however, our ancient explorers could have found abundant life by imaging Earth from the comfort of their spacecraft, much as we do with satellites today. Where organic matter was scarce before cyanobacteria, it would have become relatively plentiful after they evolved. The degradation of this organic matter drives ecosystems. More organic matter drives more active ecosystems and probably more complex ones as well. The increase in ecosystem complexity would have also resulted from the new availability of oxygen and the subsequent evolution of organisms using it. Overall, there would have been an in-crease in biological diversity as both abundant organic matter, and oxy-gen, become available in environments ranging from the land to the sea. All in all, the evolution of cyanobacteria was a (more likely "the") big deal in the history of life on Earth.

My introduction to cyanobacteria came just after I finished my PhD at Yale with Bob Berner. Bob was infinitely patient and did his best to turn a chemist into something resembling a geoscientist. Though the focus of my PhD was on how organic matter was cycled by microbes in modern marine sediments, through Bob's training and encouragement, I became fascinated with Earth history problems and in particular with the history of atmospheric oxygen. I realized already while doing my PhD that if I wished to understand something of this history, I needed to understand more about the cyanobacteria producing oxygen. This led me to a postdoctoral position with Dave Des Marais at the NASA-Ames Research center in Palo Alto, California. Dave was not only interested in the history of atmospheric oxygen, but he also had an ongoing research program studying modern cyanobacterial populations.

Dave was also the perfect postdoc mentor. We often referred to him as "Mister Wizard." This was out of great respect. Dave seemingly knew everything and could fix anything. Our fieldwork on cyanobacteria brought us to the Baja Peninsula, Mexico. The thousand-mile journey to our field site took us, in unmarked NASA vans, through impromptu and questionable checkpoints, often manned by armed militia, and finally to the Hotel El Morro in the town of Guerrero Negro. Once there, we converted one or more of our hotel rooms into labs (we tipped the cleaning staff generously when we left!) and set up our outdoor incubation tanks and our indoor electrical equipment. Our need for stable electricity was just at the edge of what El Morro could provide. On nearly every trip, Dave was busy changing fuses at the hotel's main power box (at the minimum), or more likely, rewiring some of the weak electrical connections. Once, the hotel completely lost its electricity. Dave traced this back to a naked and loose wire connection from the electrical tower supplying the hotel. Typically calm in the face of catastrophe, Dave lifted this loose connection off the ground and secured it so we could again enjoy reliable electricity. On another occasion, the cooling device that I brought to regulate the temperature of my incubations broke down. No problem. Dave rewired the water circulating system so that the device would begin to pump water when the incubation tank became too warm. By hooking the pump to a reservoir of ice happily supplied by the "depósita" next door,[4] the problem was solved.

Once we started our fieldwork, Dave's handyman's cap was replaced by his diplomat's hat. We worked as guests of the local salt company, the largest in North America, and before every field campaign, Dave would carefully explain to the attentive managers how our results might help them optimize their salt production. In those days, our research ticket into Mexico was a signed letter from the salt company director stating that our work was of great importance to their operation.

Once inside the compound, we began the long drive to our field site. In a massive complex covering around 500 km^2, seawater is sent through a long series of ponds where it is continuously evaporated away in the desert warmth and wind. In the end, salt (NaCl) precipitates from the concentrated brine. In the middle stages of this process, when the saltiness of the water is about three times that of seawater, vast expanses of spectacular cyanobacterial mats develop (plate 7). On modern Earth, cyanobacterial mats often form at high salinities, because really salty water eliminates the majority of the animals who would otherwise feed on and disrupt the cyanobacteria. On ancient Earth, after cyanobacteria evolved, but long before the evolution of grazing animals, the vast portions of the shallow seafloor bathed in sunlight would have made a suitable habitat. These regions, along with the bottoms of shallow lakes, rivers, and ponds, must have looked much like the bottoms of these present-day ponds in Baja, Mexico. On ancient Earth, cyanobacteria often formed layered structures known as stromatolites, as they sometimes also do today in certain environments (plate 8).[5] Our job was to explore the modern Baja mats, measure the activity level of the cyanobacteria, and unravel the ecology of the environment in which they live. Our ultimate goal was to understand how cyanobacterial mats on ancient Earth might have functioned.

We explored many aspects of these mats, but I will focus here on the cyanobacteria themselves. We studied a species known as *Microcoleus chthonoplastes*, which predominates in the mats. *Microcoleus chthonoplastes* is a so-called filamentous cyanobacterium, in which cells are attached end-to-end in long strings called trichomes. Anywhere from two to dozens of the trichomes are bundled together in sheaths fixed within the mats. The cyanobacterial filaments move up and down the sheaths in response to a variety of stimuli including light, oxygen, and sulfide levels. In the Baja mats, these sheaths are packed together in a network that somewhat resembles tofu in consistency. Dave Des Marais made a nice drawing of the Baja mat (fig. 4.1) based on transmission electron microscope images. In looking at this, we see that the cyanobacteria in this Baja mat (and similar to many other microbial mats in other places as well) are concentrated within a layer less than 1 mm thick. They are concentrated here, because by 1 mm depth in these mats, the visible light they use has all been absorbed.[6] As the light diminishes, cyanobacterial abundance decreases, and at the base of the layer we see a series of much smaller filaments oriented mostly horizontally in the mat. These are green in color

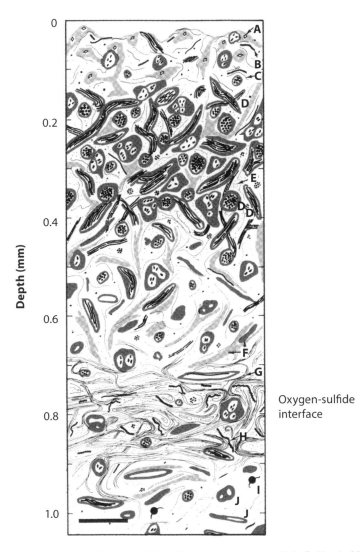

Figure 4.1. Drawing of typical microbial mat from Guerrero Negro, Baja California, Mexico, based on transmission electron microscope (TEM) observations. The oxygen-sulfide interface is located during the day at about 0.8 mm depth. The letters refer to A, diatoms; B, *Spirulina* sp. (cyanobacterium); C, *Oscillatoria* spp. (cyanobacterium); D, *Microcoleus chthonoplastes* (cyanobacterium); E, nonphotosynthetic bacteria; F, fragments of bacterial mucilage; G, *Chloroflexus* spp. (green nonsulfur bacteria, capable of anoxygenic photosynthesis); H, *Beggiatoa* spp. (nonphotosynthetic sulfide-oxidizing bacteria); I, unidentified grazer; J, abandoned cyanobacterial sheaths. Adapted from Canfield and Des Marais (1991).

and contain bacteriochlorophyll *a* and *c*. They represent a genus of anoxygenic phototrophs known as *Chloroflexus*, or informally, as green-nonsulfur bacteria. These phototrophs use wavelengths of light that penetrate below the cyanobacteria, and they use this light to oxidize the sulfide produced by sulfate reducers in the dark, deeper, oxygen-free layers of the mat. The ecology of this mat is much more complex than my simple description indicates, but this overview introduces the major players and shows, basically, how a cyanobacterial mat ecosystem can be structured.

The dimensions of this mat are so compressed that to really understand it we need tools that can probe the mat to at least a 0.1 mm resolution. Niels Peter Revsbech of Aarhus University, Denmark, faced this challenge as part of his PhD project, and in the late 1970s, he developed the first microelectrodes for measuring the distribution of oxygen, sulfide, and pH in nature. These tiny electrodes had tip diameters of mere microns (fig. 4.2) and thus the necessary fine-scale resolution. Niels Peter is a master of electrode design, and he has continued to develop these and others for ecological use. These developments have spurred major advances in microbial ecology by providing a window into the chemistry of microbial mats (and other natural ecosystems) at scales nearly matching the size of the microbes themselves. We probed the Baja mats with such electrodes, and an example of the distribution of oxygen is shown in figure 4.3. To me, it is quite amazing that all the action with oxygen takes place over a thickness of less than 2 mm. Separate your forefinger and thumb by 2 mm and just imagine: in this short distance, oxygen rises to about 4 times air saturation (that's nearly 1 bar of O_2!) and drops again to nothing. That's during the day. After the Sun goes down, the peak in oxygen concentration quickly disappears.

Not long after developing the first oxygen microsensors, Niels Peter also devised a clever way of determining rates of oxygen production in microbial mats.[7] We applied this method to our mats, and the distribution of oxygen production rates are shown together with the oxygen profile in figure 4.3. These rates of oxygen production are huge, and on a per volume basis, they are among the highest you can find anywhere in nature. Even if we integrate over depth and determine rates of oxygen production (which are equivalent to rates of organic matter production by oxygenic photosynthesis[8]) on an areal basis (this means the rate

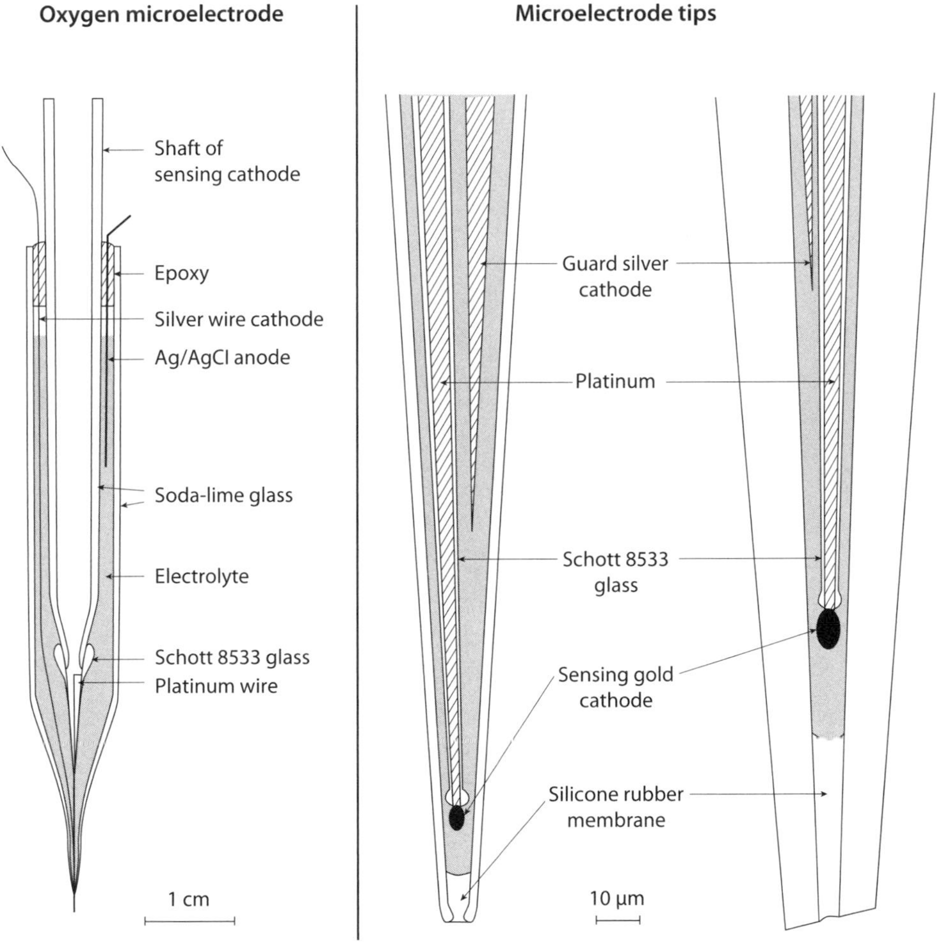

Oxygen microelectrode

Shaft of
sensing cathode

Epoxy

Silver wire cathode

Ag/AgCl anode

Soda-lime glass

Electrolyte

Schott 8533 glass
Platinum wire

1 cm

Microelectrode tips

Guard silver
cathode

Platinum

Schott 8533
glass

Sensing gold
cathode

Silicone rubber
membrane

10 μm

Figure 4.2. Diagram of a microelectrode at different magnifications as developed by Niels Peter Revsbech, and as used as a standard tool in microbial ecology. Redrawn from Revsbech (1989).

under a given area of the mat), we find that these rates are also high. For example, the depth-integrated rates of oxygen production for the data shown in figure 4.3 amount to about 20 millimoles of oxygen production per square meter of mat surface per hour. This value is much higher than the rates found through most of the global ocean, and it is matched or exceeded only in the most highly productive coastal areas. And in these mats, all of this oxygen production occurs over less than 1 mm depth! This is remarkable. Rates of oxygen production and depths of

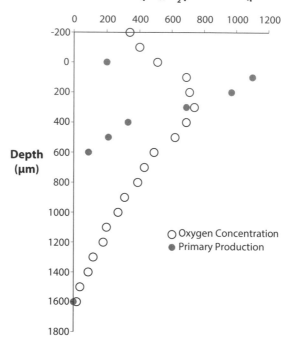

Figure 4.3. Oxygen distribution and rate of oxygen production for microbial mats in Baja, California, Mexico. Data from Canfield and Des Marais (1993).

oxygen penetration vary from mat to mat, but as a rule, cyanobacterial mats are extremely active.

Cyanobacteria are also not rare. They are well represented in microbial mats like those we studied in Baja, and if we look hard, we might convince ourselves that they are almost everywhere. We find them in the beach sands of Bornholm as described in chapter 2. But that's just a beginning. They form mats in the bottoms of lakes, ponds, and streams, and are common in soils lacking higher vegetation. Thus, they are found in particular abundance, rather paradoxically, in arid soils where water is scarce. These cyanobacteria have the remarkable ability to survive extreme drought in a dormant state, and to spring to life when water becomes available. Travel to the Mojave Desert and scoop up some sand. There will almost certainly be a reasonable population of cyanobacteria in your hand. Travel next to Uluru (formerly called Ayers Rock) in Australia. The sands and soils surrounding this great monolith will surely contain cyanobacteria, but take a close look at the rock itself. Living at, and especially just beneath, the surface of this natural sandstone sculpture are abundant populations of cyanobacteria. Travel to the Dry Valley regions of Antarctica, and you will find something similar in the giant cliffs of the Beacon Sandstone facing Lake Vanda in Wright Valley. Sandstone and granite outcrops containing cyanobacteria are found in other arid and semiarid environments ranging from North Transvaal, South Africa, to the Orinoco lowlands of Venezuela.

Let's look in the oceans. Scientists have studied marine algae for centuries. Yet, there's no mention of cyanobacteria in Claude Zobell's classic treatise *Marine Microbiology* from 1946, nor were marine cyanobacteria described in the biological oceanography textbook that I used in 1980. In fact, by the time my biological oceanography textbook was written (1977), bundle-forming cyanobacteria of the genus *Trichodesmium* had been described (more on these interesting organisms below), and other small coccoidal-shaped cyanobacteria had also been identified; but apparently their novelty was such that they had not yet entered into textbook discussions. The situation changed in 1979 when John Waterbury from the Woods Hole Oceanographic Institution, along with colleagues, described the widespread abundance of tiny marine cyanobacteria of the genus *Synechococcus*. And as it turns out, these little guys are common, except in high Arctic regions, with abundances ranging up to one million cells per liter of ocean water.

How could they have been so long overlooked? For decades, marine phytoplankton (which are photosynthetic plankton) had been collected for study in nets with mesh sizes of about 20 microns (0.02 mm). However, *Synechococcus* cells are tiny, with diameters ranging from about 0.8 to 1.5 microns. Therefore, net as many phytoplankton as you like, and *Synechococcus* cells will just wash through the net. But, John collected cells on filters with much smaller pore sizes (0.2 microns) and was thus able to isolate these tiny *Synechococcus* cells. As often occurs in scientific discoveries, however, John was not looking for cyanobacteria at all, but for something completely different. He was instead using a new technique to quantify total bacterial cell numbers by staining the cells with a fluorescent dye and counting them under a microscope as they fluoresced under blue light. A good scientist makes controls (samples that are not treated), and John saw immediately that some cells fluoresced without the added dye. Some of the photosynthetic pigments in oxygen-producing phototrophs will do this, and because these fluorescing cells were tiny, John suspected they were cyanobacteria. This turned out to be the case, and marine *Synechococcus* was discovered.

But this was not the last word on important cyanobacterial populations in the sea. At around the same time that *Synechococcus* cells were identified as an abundant species, other similar-looking cyanobacterial populations were also observed. These organisms contained a somewhat different internal structure than *Synechococcus*, and the chlorophyll pigments were also a little different; in the absence of more detailed studies, these organisms were initially classified as *Synechococcus* cells. This changed in the late 1980s when Penny Chisholm from MIT was using a relatively new technology called flow cytometry to explore the nature of photosynthetic populations in the sea. Using this technique, one can distinguish populations based on their size and on their ability to fluoresce under different wavelengths of light. Many populations, such as *Synechococcus,* generate distinctive signals, yielding an excellent tool for quantifying their population size. In applying flow cytometry, Penny was indeed able to find and quantify populations of *Synechococcus*. Another population of fluorescent cells also appeared, however, and these differed from *Synechococcus* in many important ways. First, with diameters of a mere 0.6 to 0.8 microns, these tiny cells were much smaller than typical *Synechococcus* cells. While they were clearly cyanobacteria, they also had unique pigments not found among *Synechococcus*, or any other phototrophic organism for that matter. They were also abundant, even more abundant than *Synechococcus*, at least in some places; and while *Synechococcus* tended to prefer the upper, well-lit regions of the water column, this new population preferred the deeper, darker regions. These were, indeed, the same *Synechococcus*-like cells reported earlier. After noting similarities with a group of symbiotic cyanobacteria of the genus *Prochloron*, Penny named these new cyanobacteria *Prochlorococcus*.

So, within about 10 years, our understanding of marine phototrophic communities completely changed. But there is even more to the story. These tiny cyanobacteria, both *Synechococcus* and *Prochlorococcus,* are not only abundant, but they can contribute up to 50% or more of the primary production in some places of the ocean. So, even though they are hard to see, cyanobacteria are a big deal in the carbon cycle.

And there's still more because I simply cannot help but explain about the ecology of *Prochlorococcus* in the oceans. I do this partly because it's fascinating, and partly because when thinking about the distant past, we can also imagine that ancient organisms must have lived with similar, and equally fascinating, adaptations to their environment. So to begin, *Prochlorococcus* species contain only about 2000 genes, making them among the most gene-deficient oxygen producers known. You would think more genes would be good, giving an organism more flexibility and a better chance to adapt to changing environmental conditions. *Prochlorococcus*, however, uses a different strategy. Rather than covering all eventualities with a huge genome, these cyanobacteria are lean and mean, and tune their genome to specific environmental conditions.

The story continues by recognizing that *Prochlorococcus* isn't just *Prochlorococcus*. A number of different *Prochlorococcus* strains have been isolated from the environment, and while some are better adapted to high-light conditions, others prefer low light, and others still are adapted to the particulars of nutrient availability at different depths in the ocean. These different *Prochlorococcus* ecotypes, as Penny Chisholm calls them, stratify in the marine water column, and this stratification also changes with latitude and longitude relative to their preferred light and chemical environment. An example of such stratification is shown in figure 4.4. If we begin with the distribution of *Prochlorococcus* near the equator, we see that the ecotypes adapted to high-light conditions prefer the upper water column, and lower-light ecotypes prefer the deeper waters. If we move to higher latitudes (48°N) where less light hits the sea surface, the intermediate-light ecotypes populate the surface layers and dominate overall. Here the ecotype adapted to the highest light conditions is greatly reduced in abundance, and the lowest-light adapted ecotype is almost completely missing, where, apparently, there isn't enough light for it to compete with the other ecotypes.

We can get some sense for what an organism does, and how it adapts to the environment, by looking at its genes; this is because genes represent a blueprint of all the different processes that an organism conducts. To date, the genomes have been fully sequenced for 12 different *Prochlorococcus* types. If these are compared, a total of about 1270 genes are shared in all of the different genomes. These represent the core of what *Prochlorococcus* does for a living, and they include the genes controlling

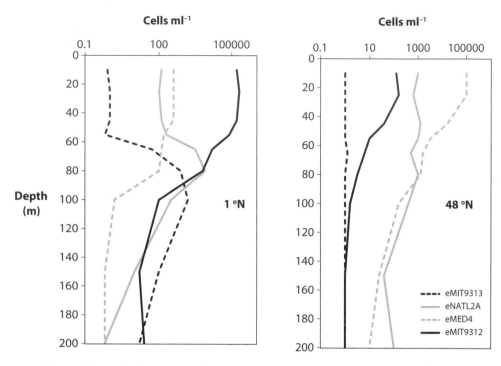

Figure 4.4. Depth distribution of different *Prochlorococcus* strains at 1°N and 48°N in the Atlantic Ocean. Redrawn from Johnson et al. (2006).

photosynthesis and carbon metabolism. Added to these are a total of almost 6000 different variable genes shared by some but not all of the different genomes. These give each different *Prochlorococcus* type its flavor and determine how each is specifically adapted to its environment. The function of many of these genes is unknown, so the specific details of environmental adaptation are still fuzzy, but the message is clear. Through gene swapping and gene evolution, *Prochlorococcus* has constructed "designer" genomes specifically adapted to the different environmental conditions found in the oceans. Who needs a big genome if you can fine-tune yours to the specifics of the environment? It is unknown how widely this strategy is employed by microbes in nature, but the recognition of *Prochlorococcus* genome flexibility by Penny and her colleagues is, in my opinion, one of the big revolutions in modern marine microbiology.

As if making oxygen isn't enough, cyanobacteria also influence the chemistry of the oceans in another critical way. To better understand this, I join a cruise onboard the *Vidal Gomez*, a rickety old American oceanographic ship recommissioned into the Chilean Navy. We are traveling due west from Iquique, Chile, to a site some 20 kilometers offshore. Dolphins are playing in the bow wake, and pelicans are searching for fish in the distance. We arrive at the study site, and the dolphins disperse; it's no fun when the ship is standing still. I relax and enjoy the view, but am startled by a school of anchovies breaking the surface by the side of the boat. As the ripples settle, I look down and think that just 100 meters below my feet the waters are completely oxygen free. We are studying one the great oxygen-minimum zones (OMZs) in the world's oceans. These are found off the western coasts of Peru and northern Chile, and to the north, off the coasts of Central America and Mexico, as well as in the Arabian Sea in the Indian Ocean. In these oxygen-free areas, microbes convert the nitrate in seawater to N_2 gas and are responsible for about one-third of all nitrate removal from the oceans. Without replenishment, nitrate, a key nutrient in algal production, would be depleted from the ocean within about five thousand years.

Luckily, replenishment occurs, and it does so in many places over the Earth surface as nitrate loss to N_2 gas is not just a marine phenomenon. The process of replenishment is known as nitrogen fixation, where many different types of microbes (more specifically prokaryotes[9]) convert N_2 gas to ammonium for use by the cell. This is a very energy-intensive and complicated process conducted by the enzyme complex known as nitrogenase. On land, nitrogen-fixing prokaryotes often live in symbiotic association with plant roots; the roots of legumes like soybeans are a good example. In aquatic environments, cyanobacteria are the main nitrogen fixers. This makes sense because cyanobacteria obtain ample energy from the Sun to run the nitrogen fixation process. Rather paradoxically, oxygen, the main product of cyanobacteria, also poisons the nitrogenase enzyme. Evolution has produced a number of very clever solutions to this apparent dilemma, and I will outline a few of these.

A number of filamentous cyanobacterial types have developed special cells called heterocysts. These are spaced along the filament at semiregular intervals, and nitrogen fixation occurs in these cells. Heterocysts contain multiple cell-wall layers that restrict the diffusion of oxygen into the cell, and while they contain photosystem I, which produces the energy to drive nitrogen fixation, they have no photosystem II. Therefore, unlike normal cyanobacterial cells, no oxygen production takes place in the heterocyst. Instead, they contain special proteins to consume oxygen from solution, providing the perfect oxygen-free environment for nitrogen fixation.

Cyanobacteria without heterocysts must find other ways to shield nitrogenase from oxygen, and there are lots of different approaches. Some cyanobacteria living in microbial mats, for example, only fix nitrogen at night when the mat environment becomes anoxic. Other cyanobacteria also restrict nitrogen fixation to the night, but they fix nitrogen in oxygenated waters. In this case, the cyanobacteria respire rapidly enough to remove oxygen from within the cell so nitrogen fixation can proceed. The most fantastic solution, I think, comes from *Trichodesmium*, which may be the most important nitrogen fixer in the oceans. *Trichodesmium* is a filamentous cyanobacterium lacking heterocysts, and it often forms impressive bundles observable by the naked eye. Indeed, large blooms of *Trichodesmium* are easily seen by satellite. Completely counter to intuition, *Trichodesmium* displays its highest rates of nitrogen fixation at midday, when light intensity is at a maximum. This would normally also be the time of day when rates of oxygen production by photosynthesis are highest because rates of photosynthesis usually increase with light intensity. Not so with *Trichodesmium*. At high light, these cyanobacteria turn photosynthesis down and use the light energy to drive oxygen-utilizing reactions in the cell.[10] In contrast, in the dim light of the morning and evening, photosynthesis is switched on, and nitrogen fixation is turned down. This strategy works when light is plentiful, which may explain why *Trichodesmium* prefers clear water and tropical latitudes (but who doesn't?).

Up to now we have explored various aspects of cyanobacterial ecology and physiology, but we have purposefully ignored plants and algae. What about them, since they also make oxygen? Indeed, could there be some relationship between the diminutive cyanobacteria that we often must struggle to see and the plants and algae that nearly define our modern world? As it turns out, the relationship is strong, and demonstrates a beautiful happenstance in the history of life. Sometime long, long ago, a cyanobacterium took up residence in a eukaryotic cell. This was a mutually beneficial (usually known as symbiotic) relationship where the eukaryote likely gained food from the cyanobacterium, and the cyanobacterium likely gained shelter within the eukaryote. Over evolutionary time, the eukaryote took control of the cyanobacterium, which subsequently lost much of its own metabolic machinery. The cyanobacterium slowly lost its identity as a separate organism and became the chloroplast of the ancestral eukaryotic algae. This fascinating idea was first developed by the Russian botanist Konstantin Sergeevich Merezhkovsky in 1905. The idea was largely lost, to be rediscovered and made famous by Lynn Margulis many decades later. It has since been proven correct with modern molecular biological techniques that clearly and cleverly showed that the chloroplast contains cyanobacterial DNA.

We owe much to our little oxygen-producing friends. They have given us oxygen to breathe, and they make sure there is plenty of nitrate in the oceans to support the great food webs of the seas. They engaged in partnerships with early eukaryotes to give us algae and eventually plants, and they can be found today populating countless different environments with fascinating adaptations to each of them. For me, it's hard to think about life on Earth without winking at the cyanobacteria and thanking them for making so much of it possible.

CHAPTER 5
What Controls Atmospheric Oxygen Concentrations?

Breathe in, breathe out, breathe in, breathe out. Nice and relaxed. We each do this perhaps 20,000 times a day and we rarely give it any thought. I think more about breathing, however, if I travel to Santa Fe, New Mexico, with an elevation of 2100 meters (7200 ft.) above sea level. Just after arrival, I'm panting after a flight of stairs, and during a short run in the hills I huff and puff much more than usual. At this altitude, atmospheric pressure is about 77% as great as at sea level, meaning that for the same breath, we only pull in 77% as much oxygen. At the top of the world, on the peak of Mount Everest in the Himalayas, the elevation is a staggering 8848 meters (29,035 ft.), and here a lungfull of air contains only about 31% as much oxygen as at sea level. Only the best trained and best adapted can survive with so little oxygen, and for only a short time at that. Most climbers make their final push up the mountain with oxygen tanks. Some don't, and many die trying. The top of Mount Everest clearly pushes the limits of what humans can endure.

So, the amount of oxygen in the air does matter. Currently, the oxygen content of the air is 21%, and we might rightfully ask why this particular concentration.[1] We might also ask whether this concentration has changed over time. We will look at the history of atmospheric oxygen in subsequent chapters, but here we will concern ourselves with the more fundamental question of why there is oxygen in the atmosphere at all. Through this discussion we will try and identify the main processes controlling the oxygen concentration.

Big deal, you might think. Any third grader knows that oxygen comes from photosynthesis. That's why we have oxygen in the air, so what's the problem? True enough, but it takes a really clever third grader to think of the following experiment. Put a plant in a closed container and watch how much oxygen accumulates during the day.[2] Write it down. Now, watch during the night. You will probably see that during the night, the plant uses nearly as much oxygen during respiration as it produced during the day.[3] We're nearly back were we started—but not exactly, and that's the point.

Look again at the measurements. The oxygen value is likely a little bit higher, and if we follow this experiment over some weeks, we might indeed see some oxygen accumulate. This is because the plant has grown. Recall that oxygen is a byproduct of photosynthesis and that oxygen production is balanced by the production of plant material. The more plant material created, the more oxygen produced. In our experiment, if ALL the plant material is used up again during respiration, then no oxygen accumulates, but if the plant grows, this growth represents unrespired plant material. Simply stated, if oxygen is not used to respire the plant material, it will accumulate in the jar. In this way, plant growth can be equated to oxygen accumulation. It's as simple as that; or is it?

Let's try the following calculation. There is currently in the atmosphere 3.7×10^{19} moles of oxygen. That's a big number. If we cooled this down to liquid, we would form a layer of liquid oxygen about 6 cm (2.5 in.) deep over the whole surface of the planet. From a combination of satellite imagery and ground-based measurements it is estimated that the net rate of primary production on Earth, which is roughly the same as the combined growth of plants, algae, and cyanobacteria, is about 8.8×10^{15} moles of carbon per year. If we compare this rate of net primary production to the mass of oxygen in the atmosphere, we calculate that the oxygen in the atmosphere could be generated in a mere 4200 years. This calculation might imply that the oxygen context of the atmosphere is unstable over rather short time intervals, and that slight imbalances between oxygen production and respiration might give us Mount Everest-like oxygen concentrations on short order, or alternatively, very high levels.

Don't worry, though, there's more to the story. Let's go back to our jar experiment. The oxygen accumulates in the jar as long as the plant is alive and growing. But, what happens when the plant dies? Like the compost heap in our back yard, all sorts of bacteria and fungi will decompose the dead plant material, using oxygen to do so. Overall, whether on the land or in the sea, it is estimated that some 99.9% of the primary production on Earth is decomposed. The remaining tiny bit is buried as unreactive organic matter in marine and freshwater sediments, which may later solidify into rock. Indeed, only the organic matter that is buried into sediments and transformed ultimately into rocks escapes reaction with oxygen. Therefore, the burial of this organic matter represents a net oxygen source to the atmosphere. So, plants and cyanobacteria produce the oxygen, but it accumulates only because some of the original photosynthetically produced organic matter is buried and preserved in sediments.[4] A lump of coal represents an oxygen source to the atmosphere, as does a barrel of crude oil, organic fossils, and all of the finely dispersed organic matter that gives the beautiful black color to the ancient shales I so like to study.[5]

There's another oxygen source that we need to consider. As we learned in chapter 2, an anaerobic microbial process called sulfate reduction respires organic matter using sulfate, and produces sulfide. This process is quite common in nature. We can find sulfate-reducing bacteria in rotten eggs, in our guts, and even on our teeth. They are most prominent, though, in relatively isolated basins like the Black Sea and the Cariaco Basin off the coast of Venezuela, in many Scandinavian fjords where water circulation is restricted, and in most marine sediments at depths where oxygen has been consumed by respiration. These sediment depths range from only a millimeter or two in environments near the continental margins to centimeters or more, if we collect mud far offshore and where the organic contents are relatively low. If there is iron around, and there usually is, the sulfide reacts with the iron, forming a mineral called pyrite (chemical formula FeS_2), which we also encountered in chapter 2. In modern sediments, this "fool's gold" is typically found as beautiful microscopic raspberry-like clusters (called framboids, from *framboise*, the French word for raspberry) some 5 to 50 microns in diameter (fig. 5.1). In ancient rocks, pyrite is often found as glistening golden cubes, but in either case, the sulfide (and the iron) bound in pyrite is

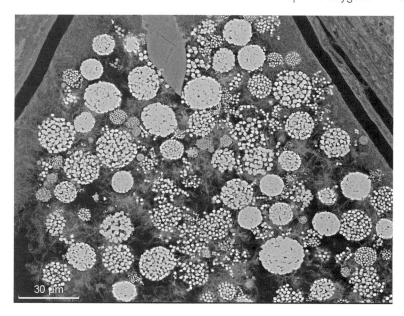

Figure 5.1. Numerous pyrite framboids in a petrographic thin section of a shale. Photo courtesy Eric Condliffe from the University of Leeds.

very oxygen sensitive, and forms into sulfate again if exposed to oxygen. Because the sulfide was formed from the reducing power of organic matter, if unreacted and buried in sediments, pyrite deposition also represents an oxygen source for the atmosphere.[6] As we will see in later chapters, organic carbon burial has been the main oxygen source to the atmosphere over the past several hundred million years, but for some intervals further back in time, pyrite burial may well have dominated as an oxygen source.

Remarkably, these controls on atmospheric oxygen concentration were first elucidated in 1845 by the French chemist and mining engineer Jacques Joseph Ebelmen; he understood these processes a mere 70 years after oxygen was discovered as an element and after its production was linked to photosynthesis, as discussed in chapter 3. Tragically, Ebelmen died at the young age of 37, but during his short distinguished career he thought deeply about which processes influence the cycling of oxygen in the atmosphere. He wrote chemical reactions that described atmospheric oxygen controls, when such representations were fairly new. Ebelmen came to the conclusion, as expressed above, that both pyrite and organic carbon burial represent oxygen sources to the atmosphere. In this way, he was able to distinguish between the short-term cycling of oxygen as defined by the photosynthesis and respiration of organisms, and its long-term geological cycling as regulated by the rock cycle. The rock cycle, as explored in chapter 1, includes the burial of sediments, the transformation of buried sediments to rocks, and the later tectonic uplift of these rocks when they become subject to weathering and erosion. The products of weathering and erosion are transported by rivers to the oceans where they once again form into sediment.

These concepts of oxygen control lay largely dormant for about 130 years until independently rediscovered in the 1970s by Bob Garrels, Ed Perry, and Dick Holland. I will have much more to say about these scientists and their contributions in upcoming chapters.

So far, we have oxygen sources under control, but what happens to all of this oxygen once liberated to the atmosphere? If you can, take a drive on a nice summer day to an outcrop of sedimentary rocks, preferably shales. If you live near a hilly region, you may find the perfect outcrop in a road cut or a river cut, or maybe just an outcrop of rocks on a hillside. If you live in a flat place like me, you may need to wait until you go on vacation. In any event, take a close look at the shale; use a magnifying lens if you have one. Notice the outermost layers. These will likely be friable and may appear bleached. You will probably see iron rust stains, and if you look closely, this rust may replace what looks like pyrite cubes or lenses of pyrite in the rock. If you have a rock hammer, dig the friable stuff away and try to find some fresh rock. The fresh rock will likely appear darker, and if you dig far enough, you should see fresh glistening pyrite. By comparing the weathered and fresh rock, you have just discovered what removes much of the oxygen from the atmosphere. Oxygen reacts with organic matter and pyrite in ancient rocks after they have been lifted back to Earth's surface. Indeed, this organic matter and pyrite, when it was first buried and formed into rock, was an oxygen source. Its eventual oxidation represents an oxygen sink and completes the cycle.

There is one other removal pathway for oxygen we should discuss. If you remember chapter 2, we discussed how chemically reduced gases like H_2, H_2S, and SO_2 entered the surface environment from volcanoes and how these probably provided energy to early ecosystems before the

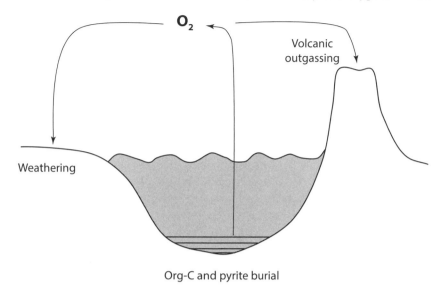

Figure 5.2. The major processes controlling the concentrations of atmospheric oxygen. Oxygen reacts with volcanic gases and organic carbon and sulfur compounds during weathering, while it is liberated to the atmosphere through organic carbon and pyrite burial in sediments. Modified from Canfield (2005).

evolution of oxygen production. After oxygen production by cyanobacteria began, these volcanic gases became oxygen sinks. Sulfide and SO_2 react with oxygen to form sulfate, whereas H_2 reacts with oxygen to form water. A simple drawing summarizing these points and showing the oxygen sources and sinks in the geological O_2 cycle is shown in figure 5.2.

Knowing all of this is critical for understanding how oxygen is cycled, but it still doesn't explain how oxygen concentrations are controlled. Indeed, can we identify processes in the oxygen cycle that actually regulate oxygen levels? The short answer is "yes." The trick is to recognize that the rates of many of the processes producing and consuming oxygen have a tendency to lead to stable oxygen levels. This may seem rather obscure, but don't worry, I'll illustrate the principle before going into more details on the specifics of how it works.

Imagine a very deep bathtub with a spigot letting water in and a single drain (fig. 5.3). The deeper the water in the tub, the faster it leaves by the drain. Turn on the tap and let water flow in. The water in the tub will reach a level where the water leaves the drain at the same rate that

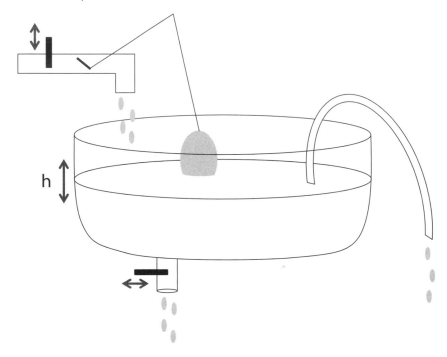

Figure 5.3. "Bathtub" metaphor for the various ways that the concentration of atmospheric oxygen is regulated; in this case, oxygen levels are expressed as the water height, h, in the bathtub. The text offers a description of how this works.

it enters. Throw a big bucket of water into the tub. The level will rise instantaneously, causing the water to leave faster. After this initial rise, the level will drop to the same point as before when input equaled output. This is known as a negative feedback, and it stabilizes the system. Take a big bucket of water out of the tub and the opposite happens. The rate of water leaving the drain slows until the original water level is again reached. The system has reached a stable state, and this state is maintained due to the action of simple feedbacks.

If we open the spigot, a new higher level of water will be reached. The water level rises and more water will exit the drain to match the new higher rate of inflow. This is another example of a stabilizing negative feedback. A higher water level will also be achieved if we constrict the drain. Conversely, open the drain or reduce the water flow, and a lower level will be reached. Now, we put in a tube and siphon water out at a constant rate. In this case, we watch the water level drop to a new level where the total rate of the water leaving by the drain and the siphon again equals the rate of input. Finally, our system has a float connected to the spigot. The float functions a bit like a toilet float (though not so sensitive), and the higher the water level, the slower the water flows from the spigot. The float helps to define a maximum water level in the tub. With this float mechanism, as the water level rises, the input rate is reduced. This is another negative feedback acting to stabilize the water level in the tub.

This simple system, of course, is a metaphor for how atmospheric oxygen is controlled, and as we shall see below and in subsequent chapters, it contains all of the principles we need to understand oxygen regulation. As with the concentration of atmospheric oxygen, this system reaches a dynamic equilibrium when inputs match outputs. A dynamic equilibrium is possible because of the negative feedbacks in the system. This system also contains a water loss mechanism, the siphon, which operates independently of any feedback. This water removal pathway influences the water level, but does not necessarily destabilize the system. In other words, a dynamic equilibrium is still possible with the siphon turned on, though at a lower water level compared to when the siphon is off.

Positive feedbacks are also possible. These are destabilizing feedbacks. An example of a positive feedback in our water control system is if the float was rigged so that the flow of water into tub increased as water level rose. With this positive feedback, the tub would overflow. As we shall see, there are possible positive feedbacks on oxygen control, but these are not the overriding controls on oxygen concentration. Oxygen concentration is stabilized by negative feedbacks. We will now look at some of these.

Let's begin with the oxygen removal pathways; we'll start with the easiest one first. This is the degassing of oxygen-reactive gases from the mantle. We are concerned in particular with gases like H_2, CO (carbon monoxide), SO_2, and H_2S, which are directly analogous to the siphon in our bathtub example. These gases are sinks for O_2 and are added to the atmosphere independently of any feedbacks in the operation of the oxygen cycle. Their rate of addition depends on tectonics and the internal churnings of Earth. As we shall see in chapter 7, these gases have played a critical role in regulating the concentrations of atmospheric oxygen during some periods of Earth history.

What about the other oxygen removal pathways, such as the oxidation of organic carbon and pyrite in sedimentary rocks, that we talked about before? These are equivalent to the drain in our bathtub example. Experiments have shown that the oxidation rates of both pyrite and organic carbon may indeed depend on atmospheric oxygen concentration, with more rapid oxidation at higher oxygen levels. This is a potential negative feedback on oxygen drawdown. If oxygen gets too low, then the rate of removal by the oxidation of organic carbon and pyrite decreases. As a first-year graduate student at Yale, I saw this point debated by Karl Turekian, a professor at Yale, and Bob Garrels,[7] whom we briefly met earlier in the chapter. Karl Turekian was a gladiator, an inspirational scientist who thrived on confrontation. Bob Garrels was a geochemistry sage, soft spoken and mild mannered. Garrels was giving the talk, and he made the point that oxygen concentration likely acted as a negative feedback on the weathering rates of organic carbon and pyrite. Karl then leaped out of his chair and savoring a fight, he roared that this couldn't be right. Weathering rates are controlled by geological uplift rates and the rates at which new rock is brought into the zone of weathering at Earth's surface. Once there, he continued, all of the organic carbon and pyrite will be oxidized before the weathered rock is carried into rivers and transported back out to sea. Feeling the temperature rise, I was tempted to climb under my desk. Garrels, on the other hand, was the picture of Buddhist calm; he smiled and said, "well, that's your opinion Karl," and continued with his talk.

In fact, Karl, as usual, made a good point. If the organic matter and pyrite is completely oxidized during weathering and transport downriver on the way to the sea, then there will be no oxygen feedback on the rates of organic matter and pyrite oxidation. Indeed, the significance of this feedback may well depend on the actual oxygen content of the atmosphere. Let's go back to the rock outcrop we discussed earlier; this is the one where we observed the weathered rock at the surface, and where the pristine rock was uncovered only after some digging. Instead of just observing the rock, let's collect samples at regular intervals from the weathered rock surface and well into the fresh rock. We take these

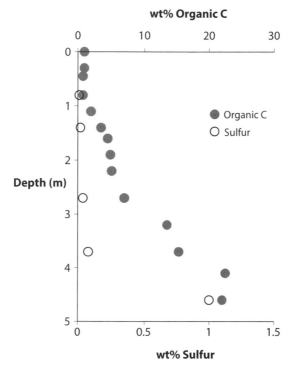

Figure 5.4. Concentrations of organic carbon and pyrite sulfur through a modern soil forming on the Woodford Shale in the Arbuckle Mountains, Murray County, Oklahoma, United States. Replotted from Petsch et al. (2000).

back to the lab and measure their pyrite and organic matter contents. Steven Petsch, now at the University of Massachusetts, did this as part of his PhD project with Bob Berner at Yale, some 10 years after I finished my PhD with Bob. Some of Steve's results are presented in figure 5.4. There are two main conclusions stemming from Steve's work. One is that pyrite is much more quickly oxidized away than organic matter. The other is that although organic matter oxidizes more slowly than pyrite, in many cases (but not all) it is almost completely oxidized in the surface layers of the outcrop.

These observations formed the basis of an elegant organic matter/pyrite oxidation model by Ed Bolton, an expert modeler and one of Bob Berner's colleagues at Yale. Ed's model is rather complex and includes a variety of parameters like atmospheric oxygen concentration, erosion rate, shale porosity (meaning basically the volume of void space, or "holes" in the shale), water flow dynamics, and pyrite and organic carbon oxidation kinetics. The results, however, are clear. Given the normal range of erosion rates and soil porosity, organic matter will completely oxidize during weathering with atmospheric oxygen levels of 25% of today's levels or higher. With lower oxygen levels, however, or with much higher erosion rates such as one might find in mountainous terrains, for example, oxygen concentration should influence the extent of organic matter oxidation. Therefore, atmospheric oxygen could have provided a negative feedback on organic matter oxidation during times in the past when oxygen levels were considerably lower than today or when erosion rates were considerably higher. A negative feedback of oxygen concentration on pyrite oxidation is also possible, but only at extremely low oxygen concentrations; this will be explored in chapter 9.

The discussion up to now has focused on the efficiency of organic matter and pyrite oxidation, but the amount of organic matter available for oxidation must also play a role. Quite simply, the more organic matter and pyrite available for oxidation, the greater the oxygen sink. The amount available will depend on several factors. One is the uplift rate of the continents as this controls the rate at which rocks become exposed to weathering. Another factor is the concentration of organic carbon and pyrite in the rocks being weathered, and this will depend, to a large degree, on which rocks become available. Bob Berner has thought deeply about this, and he has conceived of a way in which atmospheric oxygen concentrations are regulated based on this principle. Bob suggests that there is a tight coupling between the burial of organic carbon and pyrite into sediments, which represents an oxygen source, and the subsequent availability of these same sediments for weathering and oxidation, which represents an oxygen sink.

The reasoning goes like this: imagine that for some reason there is a period where sediments are formed that are very rich in organic carbon and in pyrite. The burial of these sediments would represent a strong oxygen source to the atmosphere. However, these sediments, at least many of them, should be quite prone to weathering and oxidation on relatively short time scales.[8] This is because these sediments (and the rocks they form) are subject to exposure during sea-level changes. They are also subject to uplift and erosion as a result of a number of tectonic processes including seafloor subduction and continent-continent collisions.[9] Thus, the "rapid recycling" of these organic carbon-rich and pyrite-rich rocks into the zone of weathering and oxidation would soon generate a strong oxygen sink. The end result is that the strong oxygen source represented by the initial burial of these rocks is tempered by a strong oxygen sink from their subsequent oxidation. Conversely, if there is a time of sediment deposition with low organic carbon and pyrite contents, the influence of this weak oxygen source is dampened by the weak oxygen sink generated from the rapid recycling and subsequent oxidation of this same sediment. This keeps oxygen from getting too low. The end result of rapid sediment recycling is to keep atmospheric oxygen concentrations from becoming either too high or too low. Note that this mechanism of oxygen stabilization does not represent a feedback on oxygen concentration; it doesn't depend on the oxygen content of the atmosphere. Rather, it is a stabilizing mechanism based on the coupling between sediment burial, tectonics, and weathering.

The feedbacks and controls described above relate to oxygen sinks. We can also identify various feedbacks associated with oxygen production. One important group of feedbacks relates to the influence of ocean anoxia on oxygen regulation.[10] We start this discussion by looking at sediment from the deep part of the Black Sea. The Black Sea is the largest anoxic basin in the world; below about 70 to 80 meters depth, the water is oxygen free, and if you take water from below about 120 meters depth, it stinks of sulfide, the product of sulfate reduction. Such sulfidic basins are known as euxinic basins, after the Greek name for the Black Sea, *Pontus Euxinus*.

The sediments depositing in the euxinic waters of the Black Sea are devoid of animals. They are also finely laminated and full of organic carbon and pyrite sulfur, much more than we find in ordinary mud settling around the coast of Denmark or anywhere near the beaches you might frequent on your summer holiday. Indeed, the burial rates of organic carbon and pyrite are enhanced when the water column becomes euxinic.

Before we go further, however, there is an important detail that we need to address. In fact, not all types of water-column anoxia lead to the sort of euxinic conditions we find in the Black Sea. For example, many lakes in the midwestern part of the United States where I grew up (and other places as well) become seasonally anoxic in their deep waters. However, these lakes generally have low concentrations of sulfate, so they can only support limited amounts of sulfate reduction. Therefore, when these lakes go anoxic, dissolved iron (so-called ferrous iron; there will be much more on this in chapter 7) rather than sulfide accumulates. This situation was also very common in the oceans during certain times in the geologic past, as explored in subsequent chapters. If you remember the oxygen-free waters off the coast of Chile discussed in the last chapter, neither iron nor sulfide accumulates, but these waters are instead dominated by nitrogen chemistry. Important though, is that all types of water column anoxia lead to reduced rates of organic matter decomposition (compared to when decomposition uses oxygen), leading in turn to enhanced concentrations of organic carbon in the sediments depositing from these waters.[11] Pyrite burial, in contrast, is generally enhanced only under euxinic water-column conditions (when the water column contains sulfide). In what follows, I will refer to anoxic water columns in general, which could mean any of the types just explored, unless otherwise stated.

So, with this digression behind us, we can imagine an oxygen feedback based the expansion and contraction of marine anoxic conditions. It is generally expected that the amount of anoxia in the deep waters of the ocean would increase as the concentration of atmospheric oxygen is lowered. This makes sense because with lower atmosphere oxygen levels, less oxygen will be mixed and transported into the deep ocean. However, with the Black Sea as an example, as anoxic conditions expand, the burial rates of organic carbon and pyrite sulfur (if the water-column is euxinic) should also increase. The increased burial of organic carbon and pyrite would generate a greater oxygen source to the atmosphere and thereby increase the concentration of atmospheric oxygen. This effect, however, has a limit. Elevated oxygen levels will, in turn, reduce the extent of ocean anoxia; this reduces organic carbon and pyrite burial rates and thus reduces the strength of the atmospheric oxygen source. Such a chain of events provides a nice stabilizing mechanism (negative feedback) for keeping oxygen concentrations from getting either too low or too high.

If we dig deeper, we find that ocean anoxia produces additional feedbacks on oxygen concentration. To understand these, we need to follow the fates of nitrogen and phosphorus, two of the most important nutrients whose availability controls rates of primary production in the oceans. Let's start with phosphorus. In fact, most geochemists generally view phosphorus availability as the most important factor controlling the rates the primary production in the oceans.[12] With this in mind, my fellow Yale PhD student Ellery Ingall, now at the Georgia Institute of Technology in Atlanta, made an astounding discovery during his dissertation research. He found that both modern and ancient sediments deposited in anoxic settings, like the Black Sea, contained much less phosphorus than did sediments deposited under water columns containing oxygen.[13] This observation leads to additional feedbacks on atmospheric oxygen concentrations. Therefore, in light of Ellery's observations, more expansive anoxic conditions should result in less phosphorus removal from the water column into marine sediments. We already learned that anoxic conditions enhance the burial of organic carbon and also pyrite in some cases, but with less phosphorus removal from the water into these sediments there should be more phosphorus available to fuel still higher rates of primary production. This would result in even higher rates of organic carbon burial and even higher rates of oxygen liberation to the atmosphere. Hence, the phosphorus cycle acts to enhance the negative feedback we just discussed on the role of anoxia in keeping oxygen levels from getting too low.

We see the same enhanced effect when oxygen levels rise. With higher oxygen levels, there is a reduction in the expanse of anoxic ocean water. This is accompanied by an increase in the expanse of oxygenated waters with a consequent increase in the removal rate of phosphorus from these oxygenated waters into the deposited sediments. The increase in the phosphorus removal rate will decrease the amount of phosphorus in ocean waters, which should reduce rates of primary production. Reduced rates of primary production should lead to reduced rates of organic carbon burial and, therefore, reduced rates of oxygen liberation to the atmosphere. This keeps oxygen concentrations from getting too high. These elegant ideas were first presented by Ellery, together with Philippe van Cappellen, another of my PhD student contemporaries working with Bob Berner.

With all this discussion of phosphorus, we must not forget about nitrogen. We took a brief look at the nitrogen cycle in the last chapter when we discussed how anaerobic microbes removed nitrate as N_2 gas in oxygen-minimum zones of the oceans. We also discussed how nitrogen-fixing cyanobacteria reconverted N_2 gas to ammonium, resupplying nitrogen to the biosphere. One could imagine that as ocean anoxia increased, the anaerobic processes producing N_2 gas would become more important, potentially draining biologically available nitrogen (mostly nitrate) from the sea. Therefore, in contrast to phosphorus, the expansion of anoxia in the global ocean would tend to limit the availability of N and potentially also limit primary production. The extent of this limitation would depend critically on whether nitrogen fixation could accelerate to resupply the missing nitrogen. If the resupply by nitrogen fixation perfectly balanced the nitrogen loss as N_2, then the phosphorus feedback described above would control primary production and organic carbon burial rates. However, if nitrogen fixation could not keep pace with nitrogen loss, then nitrogen concentrations would be drawn down, and nitrogen would limit primary production. In this case, the phosphorus feedback described above would be ineffectual.

My friend and colleague Paul Falkowski from Rutgers University believes that this would be the case. Indeed, in his models, ocean anoxia has the effect of decreasing rates of primary production through severe nitrogen limitation. If nitrogen did indeed limit primary production during times of extended ocean anoxia, a rather interesting positive feedback results. To see this, imagine that we have precisely the situation described above; there is widespread anoxia in the sea with severe nitrogen limitation on primary production. Now, let's imagine that for some reason atmospheric oxygen levels increase just a bit. This would cause anoxia in the oceans to decrease, and following the logic outlined above, this would also cause a decrease in rates of N_2 loss by denitrification. Because less nitrate is removed from the ocean as N_2 gas, some of the nitrogen limitation would be relieved, leading to increased rates of primary production and organic carbon burial. This would produce more oxygen, further reducing the extent of anoxia in the oceans, and so on. In principle, this positive feedback would operate until nitrogen was no longer the limiting nutrient. Then, the phosphorus feedback on atmosphere oxygen regulation would take over.

There is one last feedback we need to consider. Those of us who work with tanks of compressed gas know that pure oxygen is reactive stuff.

We keep sparks and flames well away from it, and we only use oxygen when we know there is good ventilation, or better yet, in a fume hood where excess gases are quickly removed from the room. If you have an oxygen tank at home coupled to a respiration mask, you know the rules: No smoking, no sparks, and no flames of any kind around the oxygen equipment. The basic reason for all this precaution is that burnable things burn much better in pure oxygen than in air. Burning experiments suggest that if we double the amount of atmospheric oxygen compared to today, things like trees and grass will burn much more easily. So, here comes the feedback. The idea is that if the oxygen concentration gets too high, terrestrial plants will have a difficult time establishing themselves because they will burn easily when the slightest spark is produced, for example, from lightning. This view arises from experiments conducted by Andy Watson, now of the University of East Anglia, as part of his PhD project.[14] What it means is that high oxygen levels should lead to more rampant forest fires and reduced plant growth on the land. This fire feedback, therefore, could be of importance in defining a maximum level of oxygen in the atmosphere.

Lucky for us, and for all life on Earth, there appear to be many natural processes (negative and positive feedbacks) that control and stabilize the concentration of oxygen in the atmosphere. As we shall see in subsequent chapters, different feedbacks have operated to control oxygen at different times in Earth history. Unraveling these feedbacks allows us to more fully appreciate both the interplay among biology, chemistry, and geology in shaping the chemistry of Earth's surface environment, and how this interplay has changed through time.

CHAPTER 6
The Early History of Atmospheric Oxygen:
Biological Evidence

The twelfth-century French philosopher Bernard of Chartres is quoted as saying:

> We are like dwarfs on the shoulders of giants, so that we can see more than they, and things at a greater distance, not by virtue of any sharpness of sight on our part, or any physical distinction, but because we are carried high and raised up their giant size.[1]

This sentiment has endured the centuries, and is as true today as it was 900 years ago. In this chapter we begin our discussion of the history of atmospheric oxygen through geologic time (see fig. P.1 for the geologic time scale), and one of the giants in this discussion is Vladimir Vernadsky, the Ukranian mineralogist turned geochemist and visionary thinker. Vernadsky was born in 1863 and died in 1945. He therefore lived through incredible turmoil, including two world wars and the fall of czarist Russia. In 1926 he published his magnum opus *The Biosphere*, in which he systemically explored how life works as a geological force. In this volume, his observations were keen, his reflections were deep, and his pronouncements were grand. One subject he touched upon was the history of atmospheric oxygen. He initiated this discussion by stating the following:

In all geological periods, the chemical influence of living matter on the surrounding environment has not changed significantly; the same processes of superficial weathering have functioned on the Earth's surface during this whole time, and the average chemical composition of both living matter and the Earth's crust have been approximately the same as they are today.

Thus, in looking at rocks through time, as far as Vernadsky could see, the influence of life on Earth's surface had remained about the same. This led him to conclude, somewhat later in the text:

The phenomena of superficial weathering clearly show that free oxygen played the same role in the Archean Era that it plays now.... The realm of photosynthesizers in those distant times was the source of free oxygen, the mass of which was of the same order as it is now.

When I first read these passages, I did a double take. This is the first account, of which I am aware, of the history of atmospheric oxygen through time. It's not very detailed for sure, and as we proceed in our discussions, we may not agree with Vernadsky's conclusion. But that doesn't really matter. What does matter is that he imagined the history of atmospheric oxygen was an addressable scientific problem, he determined what evidence could be used to address it, and he did so within the limits of the observations available at the time. Pretty cool, I think.

While Vernadsky is a scientific hero in Russia, he is, sadly, barely known in the West.[2] Part of the problem is that the first English translation of *The Biosphere* was finished only in 1977, and perhaps even more significantly, the post-WWII "Iron Curtain" and Cold War considerably limited scientific exchange between the Soviet Bloc countries and the West. As for myself, I first learned of Vernadsky some 10 years ago and only read *The Biosphere* rather recently. It was clear while reading it, though, that Vernadsky was speaking a language I understood. Indeed, many of his ideas fit well within our current understanding of how the biosphere and the geosphere are coupled.

How could Vernadsky sound so familiar if he had been so overlooked by western science? One explanation is that if scientific ideas are truly great, they offer unique insights into the functioning of the natural world. Therefore, science will converge on these ideas, if not now by one scientist, then later by another. We've seen an example of this in chapter 5, where Jacques Joseph Ebelmen in 1845 accurately described the geological mechanisms controlling the concentrations of atmospheric oxygen. These ideas were lost, and were independently discovered by Bob Garrels, Ed Perry, and Dick Holland some 130 years later.[3] Thus, it is possible that in the West, we may have rediscovered much of what Vernadsky understood decades ago.

It is also possible that Vernadsky's ideas have come to us by other indirect routes. In the 1950s and 1960s Russian scientists were miles ahead of western scientists in understanding the role of microbes in the chemistry of natural waters and sediments. I have not yet been able to confirm it, but given Vernadsky's status in Russian science, he must have been a strong influence on those postwar microbial ecologists. Although much of this work was published in Russian and was not readily available to western scientists, there were several key meetings between Russian and western scientists in the 1970s and 1980s, and outcomes of these encounters are presented in a number of influential volumes.[4] Perhaps, at least in part, Vernadsky came to us by this route.

Anyway, we come back to Vernadsky's reflections, and we specifically embrace the following question: How can we know anything about the history of atmospheric oxygen? We need clues of course. Vernadsky found clues in old sedimentary rocks that were once part of the ancient seafloor. As it turns out, this was an excellent idea, but it would be fair to ask how ancient mud holds clues to the levels of atmospheric oxygen in the past. Let's have a look.

If you have ever walked barefoot through a seaside mudflat, you've felt the mud squishing up through your toes, and perhaps saw bubbles forming as your feet sank into the mud. The bubbles are methane, formed by the same methanogens we met in chapter 2. In this case, they are living off of organic matter in the mud. The presence of methanogens tells us something about the chemistry of the mud and the ecology of the microbes living there. Maybe you caught a whiff of sulfide. This is formed by another group of microbes, the same sulfate reducers we met in previous chapters as they live off of organic debris. Another clue. There is something hard between your toes. You reach down and extract an empty snail shell. Yet another clue. The organic matter in the mud also holds clues. It is created from dead organisms either in the water overlying the mud or in the mud itself. If we're lucky, this organic matter can tell us something about the organisms once living in the environment. The mud particles themselves react with chemical constituents in ocean water as the particles settle to the seafloor. The mud reacts further after it's in place on the seafloor, and as the sulfide and methane accumulate. Even more clues. Overall, the mud is awash with clues. Our challenge, as we shall see as the story unfolds, is to appreciate which clues are there and to understand how they relate to the oxygen content of the atmosphere. So Vernadsky was right, the clues are in the mud, but understanding them is not always easy.

OK, so let's grab a bunch of old sedimentary rocks and get started. We go to the rock store and ask the lady at the counter for rocks from, let's say, both the shallow and deep parts of the ocean. (Why not? A comparison might be interesting.). Let's get samples from every 10 million years at ten places from around the globe (we want to ensure good coverage), from the birth of the planet 4.55 billion years ago until 2.5 billion years ago (the end of the Archean Eon, and the focus of the current chapter). That's lots of rocks, but what the heck, better too many than too few. The lady's jaw drops, and she opens the storage room door wide. We immediately see the problem. The shelves are mostly bare. As much as the shopkeeper would love to help, the rocks just aren't there. In fact, there's nothing on the shelves but a few small mineral grains from before about 4.0 billion years ago.[5] The first 500 million of years or so of Earth history is basically not represented in the rock record. We find a few rocks from 3.8 billion years ago, some more at 3.5, some at 3.2, 3.0, 2.9, 2.7, and some even younger, with rather better coverage up to 2.5 billion years ago. But, in general, the early geologic record is very poor. The problem is that the very tectonic processes we discussed in chapter 1—the subduction, mountain building, weathering, and the constant recycling of rocks at the Earth surface—all these processes that help make Earth a terrifically habitable planet also wreak havoc on the geologic record. And the older the rocks, the more likely they have been lost to weathering or otherwise been caught in and been transformed by the macerations of Earth.

We thank the lady at the rock store, take what we can get, and we go on our way. We pause under the shade of a nearby tree and see what we

$$\delta \, ^{13}C = -26 \, \text{‰} \, [PDB]$$

Figure 6.1. The letters and digits here were written with the graphite in metamorphosed sedimentary rocks from Isua, Greenland. The text is a nod to the typical isotopic composition of organic carbon in sediments. Photo kindly provided by Minik Rosing.

have. On close inspection the oldest rocks, and we don't have many of these, are also in pretty bad shape. By this I mean that many of them have been heated to high temperatures, in some cases more than once. This heating changes the original minerals in the rock to other forms through metamorphism. In a rather telling example, the original organic matter in some of our oldest examples, the ~ 3.8 billion year old rocks from Isua, Greenland, has been completely converted to graphite through heating. You can actually write with these rocks (fig. 6.1). In the rocks from Greenland, and in many of the others, fluids have also flowed through the rocks when they were deeply buried in Earth's crust, in some cases redistributing the chemical constituents.[6] My point here is not to be discouraging but rather to be realistic about the limitations of the geologic record, particularly when we are looking way back in time. Nevertheless, despite the difficulties, there are some things we can say about both the chemistry and biology of earliest Earth. We will begin by exploring early Earth biology. In this chapter, we will be particularly interested in looking for any signs of cyanobacteria. Remember, they were the first oxygen producers on Earth and were the stars of chapter 4. Without cyanobacteria, oxygen could not have accumulated into the atmosphere. We will focus in the next chapter on chemical evidence for early Earth oxygen.

We begin our quest with the ~ 3.8-billion-year-old graphite-rich rocks from Isua, Greenland. My friend and colleague Minik Rosing from the Geological Museum in Denmark (we briefly met Minik in chapter 2) has spent much of his career exploring the rocks of Greenland. In fact, Greenland holds a special place for Minik as he was born there, and his father, Jens Rosing, was a well-known Greenlandic painter, illustrator, jewelry designer, and author. After many years of fieldwork, Minik discovered these graphite-rich sedimentary rocks and immediately recognized them as good candidates for revealing the nature of early life on Earth. The sediments themselves, exposed on an outcrop not much big-

Figure 6.2. Sedimentary rocks from Isua, Greenland. Note the near vertical, black, graphite-rich layer running from top-center to lower-left. There is a similar layer just to the right. Photo kindly provided by Minik Rosing.

ger than a car, were apparently deposited in deep marine waters. The graphite-rich layers are interbedded with sediment layers known as turbidites.[7] These are formed as sediment from shallower water depths is remobilized and transported rapidly downslope where it is redeposited in deeper waters (fig. 6.2). Minik interprets the graphite layers as the background deposition of organic matter-rich particles that came from the surface waters of the ocean. This constant, gentle flux of organic matter was punctuated by the occasional rapid flow of turbidites. In this way, you get a layer-cake alteration of organic-rich sediments and turbidites. The organic matter flux to the sediments must have been reasonably high to yield enough graphite to write with. But what form of life did the organic matter come from? This is the million dollar question. Could it have been produced by cyanobacteria?

The sediments are way too cooked to offer any kind of fossil evidence of what type of organisms were present, but there is other evidence we can use. Pick up a rock, turn it in your hand, and look long and hard at the graphite. Might we find some clues here?[8] Minik did this and decided to look for evidence in the isotopes of carbon preserved in the graphite. To understand this, we recognize that in nature carbon is found with three different isotopes: carbon-12, carbon-13, and carbon-14. Most have heard of carbon-14; it's radioactive, forms in the atmosphere, and sticks around for only tens of thousands of years, so we won't worry more about it. Carbon-12 has six protons and six neutrons, whereas carbon-13 matches its six protons with seven neutrons. Therefore, carbon-13 is about 8% percent heavier than carbon-12 (13/12 =

1.083). Chemically, the two carbon isotopes are nearly identical, but not completely so, and as we shall see, these slight differences in chemical reactivity give rise to signals we can detect and interpret.

We now come back to the question of life. As explored in chapters 2 and 3, many organisms make their cells from carbon dioxide (CO_2) derived from the atmosphere or dissolved in water (for example, plants do this through photosynthesis). The organisms doing this must convert the CO_2 to organic matter, and this is accomplished by a variety of different types of biochemical reactions involving enzymes. Plants, as well as cyanobacteria, use the enzyme Rubisco. We met Rubisco in chapter 3, but what we didn't say was that Rubisco preferentially uses the carbon-12 in CO_2 over the carbon-13. This means that plants and cyanobacteria are depleted in carbon-13 (and at the same time enriched in carbon-12) compared to the distribution of carbon-12 and carbon-13 atoms in the CO_2 from which the organic matter was formed. Now, we need a language to express these ratios. Isotope ratios are typically measured on instruments known as mass spectrometers. The isotope ratio we get for our sample is reported relative to the same ratio measured on a standard with a known ratio of C-13 to C-12 atoms. We end up with small numbers when we do this, so we multiply by 1000 to give us numbers we can easily discuss. For the carbon system, we report the isotopic composition of our sample as $\delta^{13}C$.[9] The more positive the $\delta^{13}C$ value, the more enriched the sample is in C-13. Turning back to Rubisco, the enzyme preferentially selects C-12 by about 2.5% over the CO_2 in the environment. If we use the nomenclature we just learned, this means that the organic matter formed by Rubisco is 25 per mil (‰) depleted in C-13 relative to the CO_2.[10]

Minik measured the $\delta^{13}C$ of graphite from Isua and found that the graphite was depleted in C-13 by about 17 per mil (the isotopic difference between graphite and the inorganic carbon found in other sedimentary rocks at Isua), which are amounts consistent with production by cyanobacteria (fig. 6.3). Do we, then, have solid evidence for cyanobacteria as far back as 3.8 billion years ago? Sadly, no. The problem is that many types of organisms other than cyanobacteria also use Rubisco, producing a similar carbon isotope signal. So, we cannot conclude that cyanobacteria were present at Isua. But, we have pretty good evidence that life was there. The Isua organic carbon isotope signal is consistent

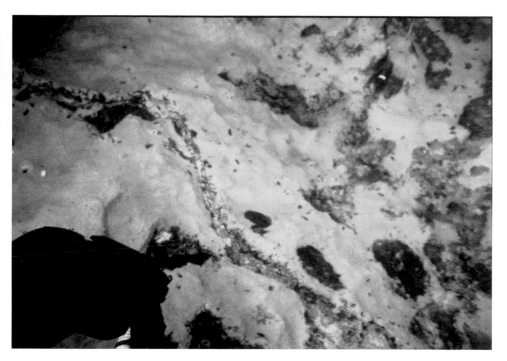

Plate 1. Mat of sulfide-oxidizing bacteria on hydrothermal sediments of the Guaymas Basin, Gulf of Mexico. Photo taken while author was onboard *Alvin*.

Plate 2. Stratified phototrophic microbial populations in beach sands on the island of Bornholm, Denmark. Photo by the author.

Plate 3. A 2.5 billion year old banded iron formation (BIF) from Dales Gorge, Western Australia. Photo by the author.

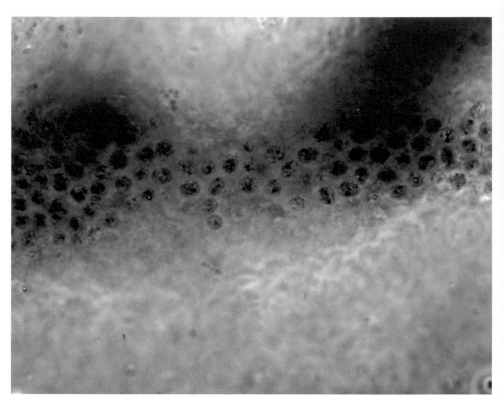

Plate 4. Natural population of purple sulfur bacteria. These specific organisms are likely involved in sulfide oxidation, but they are closely related to Fe-oxidizing phototrophic purple bacteria. Photo courtesy of Bo Thamdrup and Jakob Zopfi.

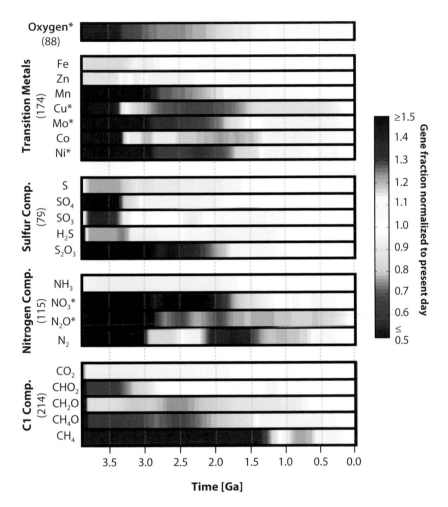

Plate 5. The evolution of the use of redox-sensitive compounds in microbial metabolism. Colors indicate the abundance of enzymes binding a specific compound relative to present day. Following this scheme, the utilization of oxygen, and a variety of nitrogen and C1 compounds (carbon compounds with only one carbon atom, including methane) evolved rather late, while manganese and sulfur-compound utilization evolved early. Slightly modified from David and Alm (2011), with permission.

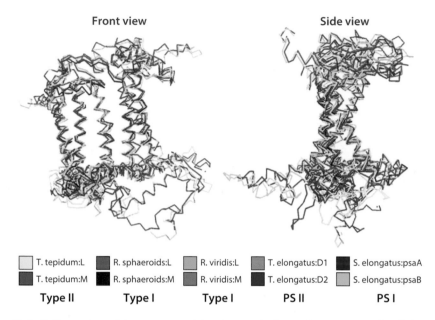

Front view   **Side view**

☐ T. tepidum:L	■ R. sphaeroids:L	☐ R. viridis:L	■ T. elongatus:D1	■ S. elongatus:psaA
■ T. tepidum:M	■ R. sphaeroids:M	■ R. viridis:M	■ T. elongatus:D2	☐ S. elongatus:psaB
Type II	**Type I**	**Type I**	**PS II**	**PS I**

Plate 6. Comparison of the structures of the photosynthetic reaction center proteins. Color version of figure 3.4.

Plate 7. Stratified cyanobacterial mat from Guerrero Negro, Baja Peninsula, Mexico. Photo by the author.

Plate 8. Modern stromatolite from Shark Bay, Australia. Photo by the author.

Plate 9. Bright red boulder of the Gaskier Formation, compared with the drab gray coloration of the adjacent boulder from the overlying Drook Formation. Photo by the author.

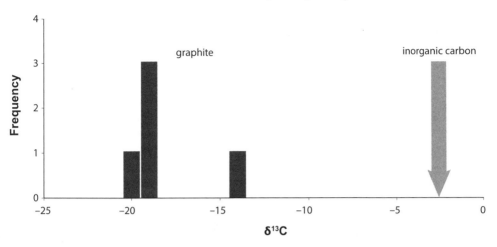

Figure 6.3. Isotopic composition of graphite and inorganic carbon from the Isua sedimentary rocks. (The frequency distribution of analyses only applies to the graphite.)

with life, as is the geologic setting. Organic matter settled from the upper waters of the ocean as we would expect from organisms fixing CO_2 in the upper ocean. Indeed, if not life, it's hard to imagine where the organic matter would have come from.[11] Though we didn't get details about the specific organisms, we have pretty good evidence that by Isua times, some form of life was making organic matter from CO_2 in the upper ocean.

Can we turn elsewhere in our search for cyanobacteria? How about fossils? Do we see anything in ancient fossils that look like cyanobacteria? We used to think so. Bill Schopf of UCLA is most famous for describing fossil-like structures from the approximately 3.5-billion-year-old Apex Chert of Western Australia. Examples of these are shown in figure 6.4. Although not terribly well preserved, many of these structures appear to consist of multiple cells arranged in filaments (so-called trichomes), and the filaments are big (by microbial standards), ranging up to over 50 microns (0.05 millimeters) in length. Bill has made several compilations of the sizes of different microbial groups, and cyanobacteria tend to be bigger than most. Indeed, they dominate in a range of sizes that match those observed for these Apex fossils. Assembling this information, Bill concluded that these fossils represent "probable" cyanobacteria; though not proof of cyanobacteria, the fossils were taken as pretty compelling evidence. I say not proof for at least two reasons.

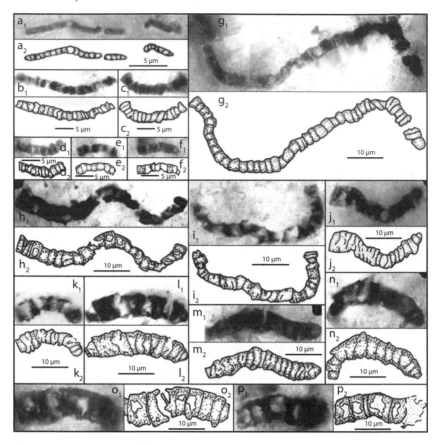

Figure 6.4. Fossils of the Apex Chert. From Schopf and Kudryavtsev (2012). Reproduced with permission.

One is that these fossils are not well preserved, and convincing cyanobacterial traits, other than size, were not observed.[12] Another issue, also known at the time, was that some noncyanobacterial organisms produce filaments of similar size and shape to those of the Apex Chert.[13] But still, despite these caveats, many (perhaps even most) supported the view that these Apex Chert fossils represented "probable" cyanobacteria, and this idea stood for a decade.

Some ten years ago, Martin Brasier from Oxford University reexamined many of Bill's fossils, which are found in specially made thin sections of rock observable under a light microscope. With improved three-dimensional imaging techniques, Martin was able to recognize branching features associated with some of Bill's original fossils, which were not recognized by Bill in his descriptions. The style of the branching observed by Martin seemed to lack obvious microbial counterparts. Martin also carefully surveyed the rocks for other possible biogenic features and found that there was plenty of organic carbon in the rock, but it ranged in size and shape from blobs to more organized structures resembling fossils. In the end, Martin concluded, quite in contrast to Bill, that the features originally interpreted as "cyanobacteria" were not cyanobacteria at all, and indeed, they were not even fossils. In Martin's view, these pseudofossils could best be explained by inorganic processes. He explained that the organic matter, mobile in these rocks as they were heated to high temperatures deep in Earth, concentrated around quartz grains and gave rise to a variety of shapes and even to fossil-like accumulations in some cases. Martin also noted that the geology of the deposits may be more complex than originally appreciated. While Bill viewed these deposits as accumulating on something like a beach, or at the mouth of a river, more recent reconstructions suggest that they were formed deep within Earth. A setting deep in Earth would not be an expected place for cyanobacteria, given their need for light, unless the particular rock pieces housing the fossils were somehow transported from a more favorable environment.

Like a game of high-stakes poker, Martin's concerns compelled a response from Bill, wherein he used even more advanced imaging techniques (Raman spectroscopy) to ascertain that indeed the walls of the "fossils" are kerogen (a kind of resistant organic matter) and that the kerogen seemed to form, at least in some cases, distinct compartments resembling cells (see fig. 6.5). Martin countered again and argued that these "cells" are merely organic carbon coating quartz grains. Bill again countered that was not the case. We have not likely heard the end of this discussion, but the original idea that these forms represent "probable" cyanobacteria has been taken off the table, even by Bill. Therefore, if the argument centers on whether these forms represent early evidence for life on Earth, the evidence might be there, but other evidence is both older and less controversial. Indeed, there is good evidence for life in the carbon isotopes preserved in the older rocks of Isua, as explored above. Furthermore, carbon isotopes in rocks of similar age to the Apex Chert seem to provide evidence for life as well.[14] And as we explored in

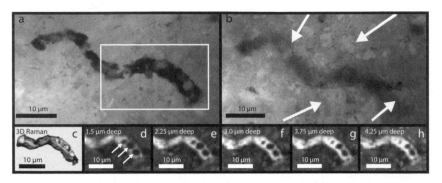

Figure 6.5. One of the Apex Chert fossils shown as an optical microscopic image on the upper left (a). Panel (c) shows a 3-D Raman image of the area in panel (a) within the rectangular box. Panels (d–h) show 2-D Raman images in different fields of focus. The upper right image (b) shows the fossils out of the focal plane but highlights that the fossils are not formed around quartz grain as shown by the arrows. From Schopf and Kudryavtsev (2012). Reproduced with permission.

chapter 2, there is also evidence for specific microbial metabolisms such as sulfate reduction and methanogenesis at that time.

So, another potentially promising insight into the antiquity of cyanobacteria bites the dust. What now? Is there anything else we might be able to look at? There are, in fact, other kinds of fossils that might help. When we think of fossils, we normally think of bones or shells, or in the case of microbes, the cells themselves. However, the cells, and particularly the cell membranes, also contain compounds that can be preserved in rocks. The cell structures may be gone, but some of their chemical constituents may persist in chemical forms known as biomarkers. Some of these biomarkers can be distinctive of certain types of organisms. This sounds pretty promising, but there are some significant issues to consider. One problem is that in really old rocks, as we are considering now, one can only expect to find miniscule amounts of biomarker molecules. Even in the best cases, these rocks have been heated sufficiently to transform most recognizable biomarkers into unrecognizable forms. Another problem, which is made doubly important because of the low amounts of biomarkers in the rock, is contamination by younger material, and this contamination can come from nearly everywhere.

The best rocks for this kind of study are obtained by drilling for fresh material buried under the surface of Earth's surface. However, normal drilling usually involves petroleum lubricants. Talk about contamination! So the drilling needs to be done with care and in the absence of organics. The whole adventure, however, could be a lost cause from the beginning if organic-rich fluids have migrated through the rocks while they were still in place underground, and this could have happened millions or even billions of years after the initial deposition. Think about how petroleum migrates to form subsurface oil reservoirs. Another problem is handling. Even if we can convince ourselves that we have collected uncontaminated samples, we must be careful not to introduce contamination as we store and process the samples, and this is harder than it sounds. Loose hair follicles, dust, pollen, fingerprints, or a myriad of nasty compounds derived from where the cores are stored; these are all sources of contamination. This work is not for the clumsy, or for the faint hearted.

Roger Summons from MIT in Boston worries incessantly about all of these problems and has one of the few labs in the world that studies biomarkers in really old rocks. His recent and very talented graduate student Jake Waldbauer, now at the University of Chicago, worked in rocks ranging in age from 2.46- to 2.67-billion-years-old from South Africa. The rocks were carefully drilled, carefully collected, and carefully handled. The rocks themselves were free of any obvious hydrocarbon migration or hydrocarbon generation, and after extraction, the small amounts of biomarker recovered had the characteristics of organic matter that was partially degraded during heating. This means that there is no evidence of recent contamination. All in all, Jake, Roger, and company have done about as good a job as possible.

What they found is extremely interesting. They uncovered a number of different types of biomarker molecules, and of interest here, they also found a variety of sterane molecules.[15] Steranes are derived from sterols, of which cholesterol is a well-known example. Sterols concentrate in the cell membrane, where they aid in enhancing the fluidity and flexibility of the membrane. Indeed, sterols are universally distributed among eukaryotic organisms,[16] including plants, animals, and fungi. Critically, as far as we know, sterols require oxygen for their synthesis in the cell. Therefore, where you find sterols, there was free oxygen, but not necessarily very much. In other work, Jake showed that sterols were synthesized by yeast (yes, yeast is a eukaryote too) with oxygen levels some 100,000 times less than one finds today in water saturated with air. Therefore, in a clever, but somewhat indirect line of reasoning, the presence of sterols means the presence of oxygen and therefore, the presence of cyanobacteria. It seems, then, that Jake (together with Roger and others involved in the study) has found evidence for cyanobacterial production of oxygen at least as far back as 2.67 billion years ago. At last, something we can hang our hat on. Well, probably. The fly in this ointment is whether, despite their great care, the biomarkers in Jake's rocks still represent contamination. Jochen Brochs, a former student of Roger's now at The Australian National University in Canberra, maintains that they most probably do. Time will certainly tell.

This quest for cyanobacteria has had some dizzying heights, some depressing lows, and it ends with a strong dash of realism. The problem is hard. The rocks aren't in good shape, and there aren't many of them. That's just the way it is. It doesn't mean, however, that people will stop looking. Perhaps a big find—maybe beautifully preserved 3.5-billion-year-old cyanobacterial cells lie somewhere underground, just out of reach, to be exposed by chance or design by some curious scientist in the future. Or maybe some other clever method will be developed to look for cyanobacteria. Indeed, that's partly what the next chapter is about. We will look at the chemistry of early Earth and see if we can use chemical approaches to uncover any signs of oxygen.

CHAPTER 7
The Early History of Atmospheric Oxygen: Geological Evidence

When I was PhD student, and just becoming interested in the history of atmospheric oxygen, it seemed that unraveling this history was the place for dreamers and hobbyists, not the place for serious scientists to waste their time. It seemed that anyone with a wacky idea could migrate into the field, deliver the idea, and then quickly retreat. The constraints were few, so even crazy ideas could find an ear.

This picture, of course, is an exaggeration. There were a few very serious scientists desperate to reveal the history of atmospheric oxygen on ancient Earth, and one of them was Dick Holland of Harvard University.[1] I met Dick's work almost immediately after starting my PhD studies and was humbled by his ability to find his way through complex data and complex problems, and to extract the simple truths (my PhD advisor Bob Berner, whom we meet more closely in chapter 11, shares the same skill). This requires a sense for pattern recognition and lateral reasoning that most of us simply don't possess. But Dick had it in spades. Dick would finish a talk and you would say: "Yes, of course, why didn't I think of that?" But the point is that you didn't, and neither did anyone else.

Dick pursued many research interests during his long career, but trying to understand the history of atmospheric oxygen was a constant through it all. Already in 1962, Dick wrote a remarkable paper entitled,

"Model for the evolution of the Earth's atmosphere." He began with a classic understatement: "A good deal of uncertainty remains concerning the composition of the atmosphere between the earliest period and the late Paleozoic."[2] He later summarized the true state of affairs: "We are therefore left with only scant evidence for the chemistry of the atmosphere during much of geologic time."

In this paper, Holland approached the history of atmospheric oxygen from an angle very different than did Vladamir Vernadsky. Vernadsky, whom we met in the last chapter, had a view that was uniformitarian and based on his observation that ancient sediments were remarkably similar to those forming today. Many new observations, however, have accumulated since Vernadsky's time, casting doubt on the likelihood of constant atmospheric oxygen levels through geologic time, and Dick's 1962 paper was a first attempt to outline a dynamic history for atmospheric oxygen levels. This paper is a classic Dick Holland contribution in many respects. In it, Dick shrewdly assembled all available evidence relevant to discussing the history of atmospheric oxygen; he then organized this evidence into a coherent picture, and attempted to quantify past oxygen levels through clever reasoning and careful modeling. Another hallmark of the Holland approach is the free acknowledgement of problems that exist and some discussion of future research that might be fruitful. In fact, Dick saw the problem so clearly that many of his observations, and much of his discussion, are still relevant today and provide a framework for the discussion to follow.

Our goal in this chapter is to explore geological and chemical evidence for the history of atmospheric oxygen on early Earth, with a focus on the Archean Eon. We start by investigating some of the evidence that Dick discussed. To do this, we put our hard hats on and travel deep into the gold mines of South Africa. Some of these mines, located in the Witwatersrand Basin near Johannesburg, have been dug to depths of over 3.9 km (2.4 miles). Without cooling, the temperatures in the mine would rise to a blistering 55°C, but with cooling machinery in place, we can safely descend and look at the rocks. These are the richest gold mines in the world, and are estimated to have produced 40% of all the gold ever mined. We look closely at the rocks and we see that they represent an ancient river deposit dated to some 2.8 to 3.1 billion years ago. The gold in these deposits was transported by strong river currents and

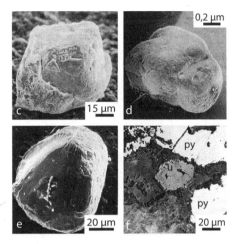

Figure 7.1. SEM images of detrital pyrites (c and d), uranitite (f), and chromite (e) from the Witwatersrand placer deposits of South Africa. Chromite ($FeCr_2O_4$) is also oxygen sensitive, although chromite oxidation seems to require the influence of microbes and intermediate oxidants, such as manganese oxides. One finds detrital chromites in rivers today. Image from Utter (1980); reproduced with permission.

was occasionally trapped among the cobbles and sands forming the riverbed.[3] This is very much like the river deposits mined 150 years ago during the great California gold rush. Our tour guide has supplied us with a Geiger counter, and tells us there is more to look for than gold. We scan among the cobbles and occasionally hear the telltale chirp. We look for the culprit with our hand lens and spot a spherical grain, now cemented by surrounding quartz; but in character and shape, it looks much like the other grains surrounding it (fig. 7.1). We are told that this is uraninite, a uranium oxide mineral (UO_2). The fact that this piece of uraninite has been worn into a spherical shape tells us that it was transported down river—hopping, bumping, and abrading away in the flowing water, much like the other sands and cobbles we find in this ancient river bed.

Uraninite, however, isn't quite like the other pieces of river sediment. We don't find it in rivers today, and that's because it reacts easily with oxygen, forming the water-soluble uranyl ion (UO_2^{2+}). Holland recognized this in his 1962 paper and used the evidence of the Witwatersrand uraninites to argue that there was at most only "trace" amounts of oxygen in the atmosphere when these ancient river sediments were deposited. Subsequently, a raging debate has developed challenging the idea

that the uraninites at Witwatersrand were indeed river sands and cobbles. Some have argued instead that the origin of the uraninite lies with
hydrothermal fluids circulating later through the rocks, and thus its
presence has no bearing on ancient oxygen levels. Many, however, still
support the original interpretation that the uraninite, at least some of it,
is part of an ancient river deposit. The shapes of the grains support this
argument, and when dated, some of the uraninite appears to be older
than the river sediments housing them. This suggests that the uraninite
was liberated from older rocks during weathering and transported downriver as originally supposed.

Luckily, river-borne uraninites are not restricted to ancient deposits
of the Witwatersrand in South Africa. Birger Rasmussen of the University of Western Australia and Roger Buick, now at the University of
Washington, described similar river-borne uraninites in sediments from
Western Australia with ages ranging from 3.25- to 2.75-billion-years-old.
Associated with these uraninites are also rounded grains of pyrite (FeS_2,
which we first met in chapter 5) and also sometimes another mineral
called siderite ($FeCO_3$). Pyrite is also very abundant in the Witwatersrand deposits, and like uraninite, both pyrite and siderite are sensitive
to oxygen. You can actually try an experiment yourself. Nearly all mineral shops will have pyrite. Buy a cheap piece and set it out in the rain
and weather. Watch and see what happens.

Other ancient river deposits in Canada (more on these in the next
chapter) and India also show signs of uraninite and pyrite transport by
rivers. Taken together then, we have evidence for uraninite and pyrite
transport in ancient rivers with ages ranging from about 3.25 to 2.45
billion years ago. What comes after about 2.45 billion years ago is the
subject of the next chapter, but the accumulated evidence supports Holland's suggestion that at most only "trace" levels of oxygen were present
in the atmosphere during the Archean Eon.

The presence of oxygen-sensitive minerals in ancient river deposits is
pretty compelling evidence for low oxygen concentrations in the early
Earth atmosphere, but you can make a stronger case if you assemble
even more evidence. To find this evidence, we turn from the land to the
sea. During the time when uraninite, pyrite, and siderite were transported as river sands, the oceans had a very different chemistry than
today. You may recall that in chapter 2 we discussed the deposition of a

rock type called banded iron formation (frequently just called BIF) in early Earth oceans as evidence of an ocean chemistry different than today's. We didn't give many details though.

Our goal now is to take a closer look at these banded iron formations and to explore in more detail what they mean for early ocean chemistry and atmospheric oxygen levels. To do this, we book a ticket to Perth, Australia, rent a car, and take a trip some 1400 kilometers north to the Karijini National Park (formerly the Hamersley Range National Park). As we drive, we pay attention to the kangaroos; they dart out in front of the car with no notice, and we keep a special lookout for road trains. We see their dust far in the distance, but we pull over to the side in good time and hold the steering wheel tight as they pass. They come fast, and as far as I can tell, they don't alter their course for anyone or anything. We also watch the gas gauge. What are listed as towns on the map in this part of Australia are gas stations. They are few and far between. Miss one, and you might not make it to the next.

The Karijini National Park contains rocks that are described geologically as belonging to the Hamersley Basin. We enter the park and drive to one of the deep gorges cut by rivers through the layers of sedimentary rock. We descend to the bottom of the gorge (plate 3) and look more closely at the rock. We are struck first by its colors: blood red, alternating with light red, gray, and white layers. When we look up and all around; this layering is apparent in every direction. We also remember that we passed kilometer after kilometer of similar rocks as we drove to the entrance of the gorge.

The red coloration in these rocks is due to a high concentration of iron minerals. These rocks are examples of the banded iron formations we've been talking about, and they reveal much about the chemistry of the oceans at the time they were formed. First, there's their wide distribution in both space and time.[4] Simply stated, they are a common rock type from early Earth. Then there's the iron; these massive amounts of iron are typically found in layers that may extend over great distances of kilometers or more. Indeed, key insights into the chemistry of the ocean and atmosphere come from understanding where all of this iron originated and how it made its way to the seafloor.

To understand this, we need to know something about the chemistry of iron (Fe). There's pure metallic iron, we all know about that, but metallic iron is rarely found at the Earth surface, so we won't consider it any further. Let's go into the backyard and check on the pyrite cube (from chapter 5) that we purposefully left out in the rain. I don't want to give too much away, but if it has been there a while, we should see reddish-brownish iron minerals forming on the surface. This is basically rust. Most iron at the Earth surface either has become, or wants to be, rust in the presence of oxygen (think about a car after 5 to 10 years of salted winter roads). Iron in this form is said to be oxidized, which in chemical terms means that, compared to the metallic form, each atom has given up three electrons. This is ferric iron and it is written as Fe^{3+}. From our perspective here, we can think of it as: oxygen + iron = rust, and rust is immobile.

My grandparents used to have a cottage on the shore of Lake Michigan, which was supplied with water from a deep well. We were told not to drink the water, which is strong encouragement for a 7-year-old boy to drink up. Anyway, the water had a particular metallic flavor that many of you probably know. I'd place a glass of freshly collected water on the table, and within minutes, brownish rust began to form on the sides of the glass. The metallic flavor was from iron in its reduced form, which from a chemical perspective, means that for every atom, it has one more electron than iron in the oxidized form. This is ferrous iron (we met ferrous iron first in chapter 2) and it is written as Fe^{2+}. Ferrous iron is soluble and quite mobile in water, but as my childhood experiment also demonstrated, ferrous iron will react with oxygen to form insoluble rust.

This is the key to understanding BIFs. Their massive scale and fine laminations mean that the iron was transported in soluble form through the ocean depths, and hence in the absence of oxygen. As this ferrous iron was mixed into the upper layers of the ocean it was oxidized, perhaps by the same phototrophic iron-oxidizing bacteria we met in chapter 2, or perhaps by small amounts of oxygen in the surface waters produced by cyanobacteria (if they were present). To be honest, we don't really know for sure how it was oxidized, but the net result was the rain of insoluble iron oxides to the ocean floor, and the subsequent formation of BIFs. We will explore this in more detail in chapter 9, but the anoxic conditions required for the deep-ocean transport of ferrous iron also require atmospheric oxygen concentrations much lower than today's values.

All of this evidence for low atmospheric oxygen levels in the Archean Eon, while compelling on its own, has largely given way to another line of reasoning that was so unexpected and so novel that nobody saw it coming. In 1999 my good friend and colleague James Farquhar was a postdoc in the lab of Mark Thiemens at the University of California San Diego. He was working late one night measuring the isotopic composition of sulfur compounds on Mark's mass spectrometer. As the results rolled off the screen, James panicked. Suddenly there was something so weird that he was sure he had damaged the machine. He shut down and went home sulking, thinking of how to break the bad news to Mark. The next morning, with a clearer head, he tried a few more samples and found that his original observations were correct. After running a series of standards, he was convinced that the machine was functioning well. After giving the data some thought, he could now tell Mark that instead of breaking the mass spectrometer, he had found a completely new way to understand the dynamics of atmospheric oxygen on early Earth.[5]

Let's look at what James did and what he saw. His objective was to measure the isotopic composition of sulfur compounds in ancient sedimentary rocks. Sulfur has four stable isotopes: ^{32}S, ^{33}S, ^{34}S, and ^{36}S.[6] In terms of natural abundance, we find most sulfur as ^{32}S, accounting for 95.02% of the total. Much less abundant is ^{34}S, accounting for 4.21 % of the total; ^{33}S and ^{36}S are only present in minor amounts, accounting for 0.75 % of the total in the case of ^{33}S, and 0.02 % in the case of ^{36}S. For decades, geologists had concerned themselves with measuring only the ratio between ^{34}S and ^{32}S. This is for two reasons. One is that these isotopes are the most abundant and therefore the easiest to measure. The other is that by standard thinking, it shouldn't matter which isotopes we look at (so why not go for the easiest). The idea is that, as we saw in chapter 6 for carbon, if some process favors one isotope over another (known as a fractionation), then all of the isotopes of the same element should behave in a predictable manner; here the degree to which a given isotope is favored (fractionation) depends on the mass of the isotope. It works like this. There is a one mass unit difference between

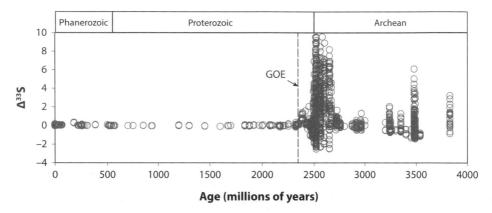

Figure 7.2. Compilation of the mass-independent sulfur isotope signal through Earth history as expressed as $\Delta^{33}S$ (see text for details). Also shown are the major eons of Earth history, and the great oxidation event, or GEO, which will be discussed in the next chapter. Data compilation kindly made available by James Farquhar.

^{32}S and ^{33}S (33 – 32 = 1), a two mass unit difference between ^{32}S and ^{34}S (34 – 32 = 2), and a four mass unit difference between ^{32}S and ^{36}S (36 – 32 = 4). In practice, if there is some process that fractionates isotopes, the effect should be about half as large between ^{32}S and ^{33}S compared to the effect between ^{32}S and ^{34}S. And, the effect between ^{32}S and ^{36}S should be about twice as large as between ^{32}S and ^{34}S. These patterns are followed through almost all biological and geological processes that fractionate sulfur isotopes. Such fractionations are known as mass dependent, because the size of the fractionation depends on the mass differences between the isotopes.

James saw something very different. In Archean rocks (and indeed rocks older than about 2.3 to 2.4 billion years), the isotopes did not behave as expected. The fractionations were not mass dependent. Rather, they deviated from these trends, generating signals that are described as mass independent. This relationship is shown in figure 7.2. Plotted on this figure is a parameter called $\Delta^{33}S$. This parameter simply designates the difference between the fractionations that are expected based on mass-dependent behavior and what is observed. If $\Delta^{33}S$ is zero, the world behaves as we expect, and indeed this is case after about 2.3 to 2.4 billion years ago. This transition to normal behavior will be the focus of the next chapter, but I think we can all agree that something weird was happening before 2.3 to 2.4 billion years ago. But what was it?

Nearly every pre-2.4-billion-year-old sedimentary rock containing sulfur was affected by it, so the process(es) producing these weird fractionations was a large-scale global phenomenon. To understand the cause of these mass-independent fractionations, we need to think about where sulfur came from and how it got into sediments. Today, rivers supply most of the sulfur to the oceans in the form of sulfate. The sulfate comes from the weathering of pyrites in the presence of oxygen and from the dissolution of sulfate-bearing rocks (most commonly gypsum), which formed sometime in the past through the evaporation of seawater. Do you remember our drive in chapter 4 with Dave Des Marais into the Mexican salt company housing those beautiful cyanobacterial mats? During that drive, we passed several ponds where gypsum was precipitating, and this occurs before the seawater is concentrated to the point of salt (NaCl) formation.

Anyway, with a good source from rivers, sulfate can accumulate in the oceans to pretty high concentrations. Take away the oxygen, however, and pyrites wouldn't oxidize to sulfate. We saw evidence for this in the Witwatersrand and related Archean river deposits, where pyrites were left unoxidized as river sands. Without a riverine sulfate source, the concentration of sulfate in the oceans becomes very low,[7] and critically, other sources of sulfate to the oceans likely become significant.

James followed this line of reasoning, and he decided that an atmospheric source of sulfur made sense because the effect is globally distributed, and he thought that volcanoes might provide a good sulfur source. Indeed, way back in 1962, Dick Holland had recognized the possible significance of a volcanic source of sulfur to the early Earth surface environment. If you measure the gases emanating from volcanoes today, you find that the major sulfur gas is sulfur dioxide, or SO_2. Hydrogen sulfide (H_2S) is also found in these gases, and whether SO_2 or H_2S dominates depends, importantly, on the chemistry of Earth's mantle at the location where the volcanic gases originate. The chemistry of the mantle, however, has likely changed little through Earth history,[8] so SO_2 was the probably also the most important volcanic sulfur species on the early Earth.

Stimulated by his initial discovery, James conducted a series of experiments in which he subjected SO_2 gas to various wavelengths of ultraviolet light, which has the appropriate energy to transform SO_2 gas into other chemical constituents. James found that he was able generate mass-independent sulfur isotope fractionations this way, and these fractionations were similar to those observed in sedimentary rocks from the Archean Eon.

Therefore, a reasonable case can be made that the mass-independent fractionations observed by James in Archean rocks were produced by the interaction between UV light and volcano-derived SO_2 gas. Now, here's the really cool part. Most of the UV light causing the mass-independent fractionation of SO_2 is absorbed today by Earth's ozone layer. Ozone is made from atmospheric oxygen. If you take away oxygen, you take away ozone, and you allow the mass-independent sulfur fractionations observed by James.

However, even if you produce a mass-independent sulfur isotope signal, it needs to be preserved. Enter again Jim Kasting, whom we met in chapter 1, and his then postdoc Alex Pavlov. They approached this problem with a complex atmospheric photochemical model and concluded that the preservation of the mass-independent sulfur isotope effect requires atmospheric oxygen levels 100,000 times less than those prevailing today (that's <0.001% of today's levels!).[9] Therefore, James's results, combined with atmospheric modeling, give us an important ancient oxygen barometer, and it says that oxygen levels on early Earth must have been really, really low. This is completely consistent with Dick Holland's original view.

One would be tempted to end the story here, but fortunately scientists are an inquisitive bunch. It's hard to tell them what to think, and they keep poking, probing, and questioning, trying to find something new. In two parts of the world, two different groups were looking closely at rocks deposited in the late Archean Eon. One group, headed by Martin Wille from the Australian National University was looking at rocks from South Africa in the age range of 2.65- to 2.5-billion-years-old. The other group, headed by Ariel Anbar from Arizona State University, was looking at 2.5-billion-year-old rocks from Western Australia (not far from the Hamersley Gorges we visited above). Let's see what they did.

These two teams both asked the same question. Could they identify some oxygen-sensitive mineral phases generating a product that could somehow be observed in the geologic record? This approach is potentially very sensitive. Suppose, for example, that we have a mineral that oxidizes halfway, or maybe only 10%, in a low-oxygen environment. Although partially oxidized, this mineral might still be preserved in a river deposit, as we explored above, and would indicate low oxygen. An oxidation product, however, would also be liberated, and if discovered by some curious scientist, it would demonstrate that although low in concentration, some oxygen was still present.

A promising element to explore is molybdenum (Mo), and both groups focused on this. On early Earth, in the absence of oxygen, molybdenum was likely present in rocks mostly as molybdenum sulfide (MoS_2) or as a minor component in pyrite. These forms of molybdenum are chemically reduced and stable in the absence of oxygen. In the presence of oxygen, however, molybdenum sulfide phases are readily oxidized to a water-soluble and mobile form, the molybdate ion (MoO_4^{2-}). This ion, once formed, is carried by rivers to the sea. Therefore, with no oxygen in the atmosphere, there is no molybdate transfer to the oceans and thus negligible molybdate in seawater. Add oxygen, though, and the concentration of molybdenum in the ocean will rise.

Both teams looked for molybdenum enrichments in the ancient sediments they explored, and both found them (fig. 7.3). The element rhenium (Re) behaves similarly to molybdenum, and both teams also found enrichments in this element as well. Therefore, each team independently discovered that oxygen reacted with compounds at the surface of Earth during a time when the mass-independent sulfur isotope record suggests very low concentrations of atmospheric oxygen.

Does this mean that James was wrong? No it doesn't. It implies, rather, that Mo and Re can be oxidized and mobilized into rivers at lower concentrations of oxygen than those required to make the mass-independent isotope effect go away. It also implies, and this is important, that cyanobacteria were around 2.5 to 2.65 billion years ago to produce the oxygen. Combined with the biomarker evidence for steranes presented in chapter 6, we now have two independent lines of evidence for cyanobacteria during a time when concentrations of atmospheric oxygen were still quite low. Ariel Anbar referred to this late Archean oxygen pulse as a "whiff" of oxygen. This name has stuck, but the question is, why only a whiff? If cyanobacteria were around, why were atmospheric oxygen concentrations so low?

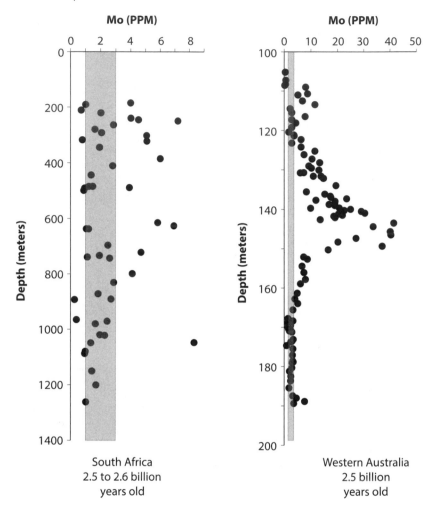

Figure 7.3. Molybdenum concentrations from rocks in South Africa and Western Australia. Values in excess of the gray box (average for the continental crust and river particles) represent enrichments and evidence for "whiffs" of oxygen during the Archean Eon.

Again, way back in 1962, Dick Holland may have had the right idea. In chapter 5 we introduced the idea that various oxygen-reactive gases spew out of volcanoes and react quickly with oxygen in the atmosphere. If the rate of introduction of these gases is fast enough, they will overwhelm the liberation rates of oxygen to the atmosphere, which as we saw in chapter 5, is controlled by the burial rates of organic carbon and pyrite sulfur. If this is the case, oxygen liberation to the atmosphere can be quite active, but oxygen will not accumulate to appreciable amounts because it is reacted out of the atmosphere by volcanic gases. As discussed in the previous chapter, we don't quite know when cyanobacteria first evolved. If they evolved well before 2.5 to 2.6 billion years ago, then volcanic outgassing of oxygen-reactive gases was clearly in excess of the oxygen liberation rates for quite some time. By 2.5 to 2.6 billion years ago, however, the rates of oxygen liberation and rates of oxygen consumption by volcanic gases seemed to be nearly balanced. Sometimes the balance shifted toward oxygen liberation, generating an oxygen whiff, and sometimes the balance shifted toward oxygen consumption, and oxygen disappeared.

The question then becomes, when and how did oxygen become more than a whiff and a permanent feature of Earth's atmosphere? This will be a focus of the next chapter.

CHAPTER 8
The Great Oxidation

The year was 1990, and I was invited to interview for a job at the University of California, Santa Barbara. Academic interviews are grueling affairs. They typically occupy two full days of discussions with faculty members, and everyone is looking for some superhuman mix of intellectual and teaching brilliance. This, combined with the right personality and interests to bind together disparate factions of the department who have not spoken in years. As usual for such an interview, I gave a departmental seminar where I discussed my work and presented some ideas as to where my work might lead in the future. Preston Cloud was there. He was a wisp of a man at that point, due to failing health, but even so, his mere presence made me nervous. I collected myself and gave a talk that may have been OK (but not great, I didn't get the job), and Cloud listened attentively through it all. After I finished, he came up, shook my hand, and said he enjoyed the talk. We exchanged a few pleasantries about Yale, where he also obtained his PhD, and then he left. I never saw him again; he was only to live another year.

Common for scientists of his generation, but unusual for scientists today, Preston Cloud's route to and through academia was indirect and varied. He was born in 1912, and after finishing high school in 1929 Cloud enlisted in the Navy, where he excelled in boxing. He was discharged in 1933 during the Great Depression, but despite the economic hardships of the time, he managed to find day work so that he could attend night school at George Washington University. Even with such a rigorous schedule, he somehow finished in four years, after which he entered Yale University as a graduate student. Once Cloud completed his PhD, he taught a year at the Missouri School of Mines but was summoned in 1941 to the wartime strategic minerals program run by the United States Geological Survey (USGS). After the war, he accepted a position as Assistant Professor at Harvard University, but left after two years to rejoin the USGS as chief paleontologist; he stayed in that position for 10 years. Cloud then moved to the University of Minnesota, and here his interest in early Earth problems was sparked. This made sense given the close proximity of the university to banded iron formations and other early Earth rocks. In 1965 he left Minnesota for the University of California Los Angles (UCLA), and three years later made his final move to the University of California Santa Barbara (UCSB), where I met him near the end of his life.

Preston Cloud's convoluted career path, and all his varied appointments, responsibilities, and experiences, no doubt provided him the breadth and depth of understanding needed for some truly BIG thinking. And think big he did, on the scale of Vernadsky, in my opinion. Indeed, in many ways, both Vernadsky and Cloud shared similar visions. The interface between biology and geology was central to both of them. However, Vernadsky's prime concern was to understand how life worked as a geological force, his interest in Earth history was secondary. For Cloud, Earth history was primary, and he was especially interested in unraveling the relationship between the evolution of life and the chemical evolution of Earth's surface environment. In 1968 he published his first "Big Think" paper with the rather academic title "Atmospheric and hydrospheric evolution on the primitive Earth." He developed his thoughts further with a 1972 paper owning one of the best titles ever in earth sciences: "A working model of the primitive Earth." With a title like that, the paper better be good, and this one was no disappointment.

Cloud's thesis was that the histories of Earth's biological and chemical evolution are intertwined. There wasn't much evidence back then to support this idea, but as a great scientist, Cloud was able to see patterns and make connections with the limited information he had. We will not just now journey through all of Earth history, as Cloud did. Rather, we will concentrate on a particular part of this history.

We start where we left off in the last chapter. That is, with evidence for low-oxygen conditions dominating through the Archean Eon (punctuated by some apparent "whiffs" of oxygen near the end). Preston Cloud, like Dick Holland before him, came to this same conclusion (without the whiffs though). But, Cloud went a step further. If atmospheric oxygen concentrations were low in the Archean, Cloud asked, when did they begin to rise? To answer this question, Cloud appealed to his experience scrambling through the rocks just north of Lake Huron in southern Ontario while working at the University of Minnesota (fig. 8.1). The rocks here are part of a large sequence of geologic formations known as the Huronian Supergroup, which range in age from about 2.5 to 2.2 billion years and thus span the age of interest. Cloud noted that the older of these rocks, with ages from about 2.4 to 2.5 billion years,[1] contained ancient river deposits with detrital uraninite and pyrite, similar to the rocks from South Africa we discussed in the last chapter. However, walking up into younger rocks, the uraninite and pyrite disappeared as did any sign of banded iron formations (BIFs).

Cloud also noted a particularly intense red staining in some of the sandstones found higher in the sequence (see fig. 8.1), above where detrital uraninite and pyrite were last observed. These rocks are known as red beds, and the red coloration reflects iron in its oxidized form. However, these rocks are very different from the BIFs we saw earlier. In many cases, red beds are formed on land or in shallow waters, and while the red coloration is prominent, these red layers are not supercharged with iron, unlike those found in BIFs.[2]

Piecing all this evidence together, Preston Cloud first, and Dick Holland later, argued for a substantial increase in atmospheric oxygen concentrations around 2.3 to 2.4 billion years ago. The uraninite and pyrite disappeared as they were completely oxidized away in elevated levels of atmospheric oxygen. Furthermore, in Cloud's view, BIFs stopped depositing as the increased levels of oxygen swept into the deep ocean, oxidizing the ferrous iron away to rust. The red beds, in contrast, formed as a direct result of weathering on land in an oxygenated atmosphere. Holland dubbed this transition to higher oxygen levels the "great oxidation event," or GOE for short. The GOE represents a quantum shift in the oxygen content of the atmosphere. This is a big deal.

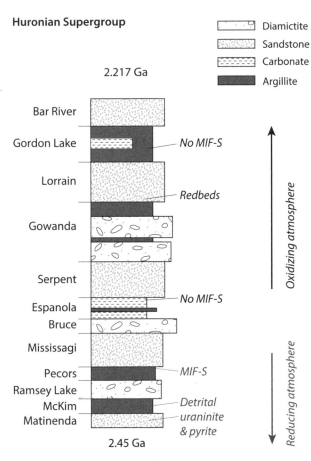

Figure 8.1. General stratigraphy of the Huronian Supergroup. Major events in the history of Earth-surface oxygenation are also shown. Note that oxygen indicators suggest the GOE (great oxidation event) occurred after the second glacial event (as represented by the Bruce diamictite; a diamictite is a glacial deposit). The MIF-S signal refers to the mass independent sulfur isotope signal, as discussed in chapter 7. Figure borrowed from Sekine et al. (2011) with slight modification. Reproduced with permission.

This is also a great story, but one difficulty with the evidence compiled by Cloud and Holland is that rocks containing detrital uraninites and pyrites, as well as those with continental red beds, are not continuously represented in the geologic record. Therefore, our resolution of the timing of the GOE from these indicators is rather poor. There is, however, another approach which will let us much more precisely explore the timing of the GOE. To follow this idea, we return to James

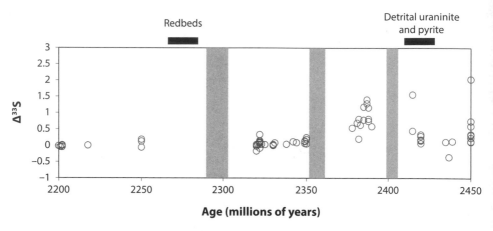

Figure 8.2. Changes in the mass independent (MIF) sulfur isotope signal through the Huronian Supergroup. Also shown are the stratigraphic levels where red beds and detrital uraninites and pyrites are found. Gray areas mark glaciations. Sulfur isotope data kindly provided by James Farquhar.

Farquhar's sulfur isotopes. Recall from the last chapter that the mass-independent distribution of sulfur isotopes was the rule in the Archean Eon, and that these fractionations were likely formed under very low concentrations of atmospheric oxygen. One of the beautiful aspects of sulfur isotopes is that most marine rocks have some type of sulfur species that we can analyze. This will usually be pyrite, but in some cases it can also be sulfate minerals. Marine rocks are almost continuously represented in the time around the GOE, so sulfur isotopes from the sulfur species associated with these rocks should yield a far more complete oxygen history.

So, let's return to the rocks of southern Ontario and measure sulfur isotopes. This, in fact, was done by Dominic Papineau of Boston College and also by James Farquhar and his group. In addition, James and his colleagues analyzed rocks from South Africa of similar age, from the so-called Transvaal Supergroup. Taken together, the results show that the mass-independent sulfur isotope signal switches to the "normal" mass-dependent signal between the last of the uraninites and the first of the red beds (figs. 8.1 and 8.2). Talk about convergence! This transition marks the GOE, at least as it influenced the mass-independent sulfur isotope signal. Our best understanding, then, is that about 2.3 to 2.35 billion years ago, oxygen concentrations rose to greater than 0.001%

of present levels, and indeed, to levels probably much greater than this. There you have it. But not so fast; some important questions still remain. For example, we still haven't explained what caused the GOE. Nor have we discussed how high oxygen concentrations actually rose during the GOE. Also, was there any obvious influence on biology? We'll focus on the first of these questions in this chapter, the second of them in the next, and the last in chapter 10.

So what caused the GOE? To be honest, lots of ideas have been proposed, and I will not discuss them all. I will, however, try to give an overview of the most promising avenues of thought. The null hypothesis, and the simplest, is that the GOE represents the evolution of cyanobacteria. Simple as that. This idea is promoted most strongly by Joe Kirschvink of the California Institute of Technology. Joe is well known for thinking outside the box and he has contributed some of the most creative ideas in modern Earth Science.[3] In this GOE debate, Joe plays devil's advocate and specifically acknowledges that he holds the "skeptical" viewpoint. He approaches the problem with the following question: "When does geologic evidence demand the presence of oxygenic photosynthesis?" He takes on nearly all of the evidence we have discussed so far for cyanobacteria before the GOE, but he invests most of his energy on the sterane evidence for an oxygen-containing environment that we explored in chapter 6.

Recall that as far as we know, steranes have an absolute requirement for oxygen in their synthesis, and therefore, finding them is a good sign of oxygen in the ancient environment. We also noted that contamination is a big concern. Joe raises this flag too, but his major argument is that while the known biochemical pathway of sterane formation has steps requiring oxygen, this was not necessarily the case through all of Earth history. He points out that the aerobic steps in sterane synthesis may have potential anaerobic counterparts that did not require oxygen. No living organism is known to conduct the anaerobic (oxygen-free) synthesis of steranes, but Joe argues that this is because the anaerobic pathways were replaced by oxygen-requiring pathways when oxygen became available. Joe's hypothesis could be correct, but it is difficult to defend without evidence. Perhaps it's a matter of style, but I prefer to side with explanations of the geologic record that rely on well-documented pathways and processes. Having said this, I appreciate those who search for nontraditional explanations to "standard wisdom." When Joe, or someone else, finds an anaerobic sterane synthesis pathway, I will happily change my views.

There's also the issue of the oxygen "whiffs" that we discussed in the last chapter. Here, in my opinion, Joe sidesteps the evidence. He doesn't really explain how molybdenum can be liberated from the continents and into the oceans in the absence of oxygen. Also, why are the molybdenum enrichments associated with other evidence for oxidative weathering on the continents like enrichments in rhenium? It seems to me that the best way to explain the geologic evidence is to accept that cyanobacteria evolved before the GOE, as we discussed in the last chapter. Therefore, we are challenged to look to other reasons for the rise of oxygen at the GOE.

Maybe we can gain some additional insight if we look again at the controls of oxygen as explored in chapter 5. If you recall, oxygen is liberated to the atmosphere, ultimately, from the burial of pyrite sulfur and organic carbon in sediments. Perhaps we can see some evidence for this in the geologic record? The sulfur story is interesting, and we will take it up in the next chapter, but take my word for it here, there is no evidence that a massive burial of pyrite caused the GOE. What about carbon then? To explore this we need to return to isotopes. The basic story is as follows. There are two major chemical forms of carbon in nature. There's so-called inorganic carbon, like the CO_2 in the atmosphere and the bicarbonate ion (HCO_3^-) in natural waters,[4] and then there's organic carbon, the stuff of life. Inorganic carbon enters the ocean mostly from rivers and mostly in the bicarbonate form and it leaves the ocean as either organic carbon, the remains of life, or as some type of calcium carbonate mineral; think of shells, corals, and limestone. Quite simply, what comes into the oceans must ultimately leave. Carbon comes into the oceans as inorganic carbon and it leaves as either organic or inorganic carbon.

Now here's where the isotopes come in. Inorganic carbon comes into the ocean with a ratio of C-13 to C-12 atoms that we usually assume hasn't changed much through time. As we explored in chapter 6, the organic carbon produced by carbon-fixing organisms like cyanobacteria and algae (these are the main primary producers of organic matter today in the sea) contains more C-12 than the inorganic carbon from

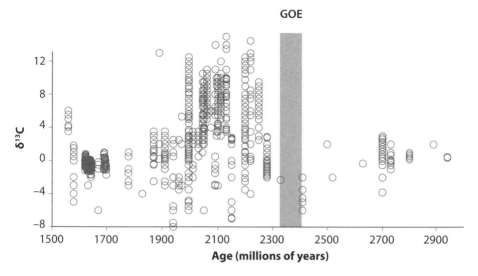

Figure 8.3. Isotopic composition of inorganic carbon showing the Lomagundi Isotope Event spanning from about 1950 to 2300 million years ago. Also shown is the GOE.

which it is formed. This means that the inorganic carbon remaining has less C-12, or in other words, it becomes enriched in C-13. The more organic carbon we remove from the oceans, the more the inorganic carbon remaining in the oceans will become enriched in C-13. We can see right away that if we have extra-high rates of organic matter removal from the ocean, we should generate extra-large C-13 enrichments in marine inorganic carbon.

We can measure the ratio of C-13 to C-12 atoms in organic matter from old rocks, as Minik Rosing did in rocks from Isua, Greenland (chapter 6), and since inorganic carbon is also removed as limestones (and shells after the evolution of animals), we can measure the ratio of C-13 to C-12 in inorganic carbon too. Thus, we can assemble a record of the C-13 to C-12 ratio of both organic carbon and inorganic carbon through time. If you remember from chapter 6, we generally discuss these carbon isotope ratios as $\delta^{13}C$ values, as we will here.

Now let's look at the data (fig. 8.3) and focus on the time of the GOE (remember, this was around 2.30 to 2.35 billion years ago). Indeed, $\delta^{13}C$ of inorganic carbon becomes highly elevated around this time, and this period of elevated values has been dubbed the Lomagundi isotope excursion. By all appearances, it is the biggest carbon isotope excursion in Earth history. This excursion was first fully appreciated in 1996 by Dick Holland and his colleague Juha Karhu from the University of Helsinki.[5] They viewed the excursion, and its associated burial pulse of organic carbon, as the source of oxygen driving the GOE. Problem apparently solved. However, if you look again carefully at the graph, you can see that things don't quite add up. Recent and better dating now puts the Lomagundi isotope excursion after, rather than during, the GOE. Rats, it made so much sense. We are forced to look for another cause.

To introduce the next possible cause we start with those fleeting whiffs of oxygen from the late Archean Eon, as discussed in the last chapter. During these whiffs, it seems as if the atmosphere experienced periodic pulses of oxygen, only to see them vanish again. We also offered in the last chapter a tentative explanation for the whiffs, suggesting that at this time in Earth history, the flux of reducing gases from the mantle was close to the flux of oxygen liberation from organic carbon and pyrite burial. Most of the time the volcanic flux was in excess, but occasionally, the balance tipped toward an excess in oxygen liberation generating a whiff of oxygen to the atmosphere.

Let's pursue this line of logic, but to do it right, we need to start way back, toward the beginning of Earth time. Indeed, we need to go back to before the beginning of the rock record, to a time when we can only use our wits and make our best guesses. What we want to know is the rate at which oxygen-reactive gases, mainly hydrogen (H_2), spewed out of volcanoes when Earth was really young.

How does one even hazard a guess? Well, let's start with today. We have some idea of how much hydrogen gas comes out of volcanoes, at least within a factor of probably 2 to 3. The degassing rate of hydrogen will depend on the chemistry of the mantle, but as we touched upon in the last chapter, this has probably not changed too much through most of Earth history, so we won't worry more about that. The hydrogen degassing rate should also depend on the rate at which the viscous material in the mantle convects or mixes. In chapter 1 we explored how this process of mantle convection drives plate tectonics, which in turn drives the recycling of materials at the Earth surface and basically allows Earth to be a pleasant place for life.

So, the question becomes, how has the convection rate of the mantle changed through time? The convection rate is driven, to a first approximation, by the temperature gradient from the interior to the surface of Earth; the higher the gradient, the faster the convection. (You can see how this works with a simple home experiment if you follow this endnote.[6]) The gradient in temperature, in turn, depends on the rate at which heat is generated in Earth. The middle of Earth is estimated to be a toasty 5500°C (although this is not known precisely). Some of the high temperature is related to the tremendous heat generated as Earth was first accreted from smaller "planetesimals" way back in time.[7] Indeed, in the later stages of this process, a massive object, something like the size of Mars, is believed to have struck Earth, spitting off the Moon as a result. This collision would have reduced Earth to a molten mass.

Heat is also generated through the radioactive decay of elements within the mantle and core. Because radioactive isotopes decay into nonradioactive chemical forms, there has been a steady decrease in the radioactivity of Earth through time, with a concomitant decrease in heat production.

All in all, the loss of heat from the early formation of Earth, as well as a reduction in the heat produced by radioactive decay, should have resulted in a cooling of the planet's interior through time. This, in turn, should have led to slower rates of mantle convection. Therefore, one can make the case that the rate of H_2 release has decreased through time in the face of slower convection. The rate of decrease in the H_2 flux through time is of extreme importance here, but unfortunately, this rate is still pretty much guesswork. If we assume, for example, that the H_2 release rate has decreased linearly with the flow of heat from the mantle, we make one set of predictions, and poor ones at that, because the history of heat flow through time is not that well known. If the H_2 release rate decreased in a different proportion to heat flow, we make another set of predictions. Quite frankly, we lack the insights to judge which rate of decrease is most appropriate, but some reduction in the H_2 flux with time to the surface environment is reasonably certain.

Despite these difficulties, many researchers, including myself, have tried to play this game. Indeed, we must if the goal is to understand the evolution of Earth surface chemistry through time, and the connection

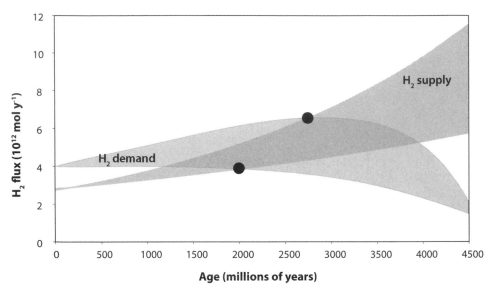

Figure 8.4. Dick Holland's calculations of the hydrogen flux from the mantle versus rates of oxygen liberation to the atmosphere. Black dots indicate crossing points for two different model scenarios. See text for details. Data taken from Holland (2009).

between the surface environment and the interior of Earth is just too important to ignore. Dick Holland also recognized the importance of this connection and made his stab at predicting the history of H_2 flux from Earth's interior through time. This estimate is shown in figure 8.4 with two lines that represent Dick's view on the range of uncertainty in the calculation.

Dick also took a stab at predicting the potential rates of oxygen production through organic carbon and pyrite burial through time.[8] I will not go into the details of this calculation here,[9] but the results are also presented in figure 8.4. In reading this graph, focus on the two black dots, not all of the overlapping grey. These dots should be viewed as the extremes of the calculation. What we see is pretty illuminating. Dick's modeling suggests that sometime between about 2.0 and 2.7 billion years ago, rates of oxygen liberation first exceeded rates of oxygen demand. This figure provides a graphic explanation for the cause of the GOE.

Is this really the way it happened? I'll stick my neck out and say, "I think so," at least in broad detail. This just makes too much sense. To be fair, others have had similar ideas, notably Jim Kasting (from chapter 1) and Lee Kump, both from Penn State University, and Dave Catling from the University of Washington. They have all described, like Dick, how the GOE likely represents a point in time when oxygen liberation exceeded oxygen demand, and they, among others, have each presented clever (but different) models to support this idea. I particularly like Dick's model though, due to its simplicity and for the fact that the processes he prescribes make good geological sense.

So, Preston Cloud had it right. The geologic record demonstrates that around 2.3 billion years ago the oxygen content of Earth's atmosphere increased dramatically. Since cyanobacteria likely evolved much earlier, it does not appear that a well-oxygenated atmosphere is a necessary or immediate consequence of the activities of oxygen-producing organisms. Atmospheric chemistry is a slave to the dynamics of the mantle, as the interior and exterior of the planet are connected in a profound way. Indeed, it took half of Earth's history for the mantle to quiet to point where oxygen could accumulate. This, however, represented a watershed, a tipping point if you will, where the chemistry of Earth's surface was forever altered. In the next chapter we will see what came next.

CHAPTER 9
Earth's Middle Ages: What Came after the GOE

Isn't it everyone's dream to travel in a time machine? Well, maybe not everyone's, but most geologists I know would love to get their hands on one. Of relevance for the story here, we could directly test with a time machine if our ideas about the history of atmospheric oxygen are correct. We've pieced this story together from our reading of the geologic record, but as mentioned before, this record gives an imperfect view of the past. We pick up a rock, a piece of ancient sediment that was once mud on the seafloor; maybe it's been heated, altering the minerals or the organic matter, or maybe much later fluids have penetrated the rock, altering its chemistry. The rock represents the sum of all the processes that have influenced it during its long history on Earth, and not just those we are most interested in.

Also, we can only surmise how the signals we are interested in were frozen in place within the rock. For example, we understand from modern environments how molybdenum finds its way into the sediment (we used molybdenum in chapter 7 to record "whiffs" of oxygen into the atmosphere). We know this because we can measure the concentration of molybdenum in the water overlying sediments today. We can trace its path from here and into sediments, and thus understand which processes act to control this path. We can make budgets and test our understanding with models and experiments. Indeed, this kind of work forms the basis for our reading of the geologic record. But, when we look at the rock we have just the concentration of molybdenum in the rock and maybe some other chemical species to give us some clues to the chemistry of ancient waters. We don't have the water from which the molybdenum was derived. Therefore, in the end, we can only estimate its concentration in the ancient oceans. This story is more or less the same for all of the indicators we use to understand the chemistry of the ancient ocean. We constantly get better at reading the geologic record, but the clues are rather few, and our interpretations are subject to big uncertainties. We could answer lots of questions with a time machine.

Our topic at hand is the aftermath of the GOE. In fact, to understand what came after the GOE, we need to go back and look in even more detail at what came before. I know we talked about the GOE in the last chapter, and what came before in chapters 6 and 7, but we need to look for small things that we may have overlooked. Things that may be difficult to determine from our reading of the geologic record. So, let's use our time machine and travel to the time just before the GOE. We put on our rubber boots (and an oxygen mask for good measure) and wade through ancient rivers. We're looking for pyrite. As revealed in chapter 7, before the GOE some pyrite settled as pebbles, unoxidized, into ancient riverbeds. But, very few of these rivers have survived, and we want to determine if pyrite transport in rivers was common or rare. Could some of the pyrite have been oxidized? Did only the big pieces of pyrite survive oxidation, while the smaller ones oxidized away? If we could go back in time, we could follow the pyrite from its origin in rocks undergoing weathering on land to its final resting place.

This would allow us to decide how the pyrite was cycled and recycled. We could see how much of the pyrite liberated during weathering on land was transported through rivers, into the sea, and back to sediments again, with these sediments themselves ultimately turning back to stone. If pyrite recycled this way, without oxidation in a low-oxygen atmosphere, then it would accumulate to higher and higher concentrations in sedimentary rocks as volcanic gases continued to deliver more and more sulfur to the surface of Earth. Such cycling would likely be true for other oxygen-sensitive minerals as well. This sequence of events makes good sense in a very low-oxygen atmosphere, but was this the way it happened?

We could also look at the fate of organic matter as it was weathered from ancient shales because these rocks house most of the organic matter in the geologic record. Going back in time, I would bring a small drill. With it, I would collect samples from the surface of shale in contact with rain and weather to deep within into the rock where weathering fluids had yet to reach. This would allow me to see how the organic matter was affected as the weathering fluids penetrated and transformed the rock.

In the modern world, as revealed by Steven Petsch and discussed in chapter 5, organic matter concentrations are universally lowest near the upper surface of a shale as the shale undergoes oxidative weathering (fig. 5.3). This oxidation requires oxygen, but it is not super fast, and some organic matter remains, even at the upper surface that is exposed to oxygen the longest.[1] So, what happens when oxygen concentrations are really low, as they were before the GOE? I would expect that organic matter escaped oxidation as the ancient shales were lifted into the weathering environment. If this was the case, then organic matter like pyrite might also have been largely recycled from rock to sediment to rock again, accumulating to higher and higher concentrations as more CO_2 entered the surface environment from volcanoes.[2] Therefore, the time machine will have told us something very important about the dynamics of oxygen-sensitive species before the GOE. In the absence of a time machine, we can only make reasonable suppositions and hope they are correct. We suppose, therefore, that before the GOE (and outside of times of oxygen whiffs as discussed in the last chapter), large proportions of the pyrite, organic matter, and many other oxygen-sensitive species brought into the weathering environment escaped oxidation and were cycled around and around from rock to sediment to rock again.[3]

Now comes the oxygen. An increase in atmospheric oxygen levels would have caused efficient oxidation of pyrite to sulfuric acid,[4] and organic matter to CO_2 with the potential liberation of any nutrients tied up with the organic matter. One might view this as the geochemical equivalent of stirring up a wasps nest! Dick Holland was, as far as I know, the first to speculate on what happened next:

> The large positive $\delta^{13}C$ excursion in carbonate sediments between ca. 2.22 and 2.06 Ga suggests that during this period, PO_4^{-3} was exceptionally available for photosynthesis and carbon burial. Several changes probably contributed to this greater availability. As

the O_2 content of the atmosphere rose, the oxidation of FeS_2 during weathering must have increased dramatically. The H_2SO_4 generated during FeS_2 oxidation must have increased the total rate of chemical weathering and, hence, the rate at which PO_4^{-3} was delivered to the oceans. At the same time, H_2S and HS^- in the near-surface oceans were probably oxidized to SO_4^{-2}. This process lowered the pH of river and seawater and perhaps made PO_4^{-3} more available for photosynthesis.

Thus, Holland argues that the acid generated during pyrite oxidation in response to the GOE would have enhanced the weathering process. It's well known that most rocks will dissolve faster under acidic conditions, but important here are those specific minerals within the rocks containing phosphorus. An important mineral in this group is apatite $[Ca_5(PO_4)_3(OH,F,Cl)]$, basically the same stuff as our teeth and bones. As with our teeth when overly exposed to carbonated beverages, the GOE, in the view of Dick Holland, produced massive apatite decay on the Earth surface. This, in turn, liberated enormous quantities of phosphorus to the oceans. This huge phosphorus input stimulated primary production resulting in high rates of organic matter burial that produced the super-sized Lomagundi carbon isotope excursion we discussed in the last chapter. In a rather unusual inversion of cause and effect, the Lomagundi excursion, which was once thought to be the cause of the GOE, now becomes a consequence.

Also, the organic matter accumulating in rocks up to the GOE will have been accompanied by key nutrients like nitrogen and phosphorus. The oxidation of this organic matter after the GOE would have released these nutrients as well, providing another source of phosphorus to the oceans. Maybe the liberation of this phosphorus was also aided by the acid generated during the oxidation of pyrite in organic-rich rocks. Overall, I rather like the idea that the accelerated liberation of phosphorus accumulated in organic-rich rocks after the GOE helped trigger the Lomagundi excursion. I like it because it provides a reason for the start of the Lomagundi as well as a reason for its end.

You might reasonably ask why we need a reason for the end. This is because of the operation of the rock cycle. We already learned in previous chapters that sadly, the geologic record of really old rocks is poor.

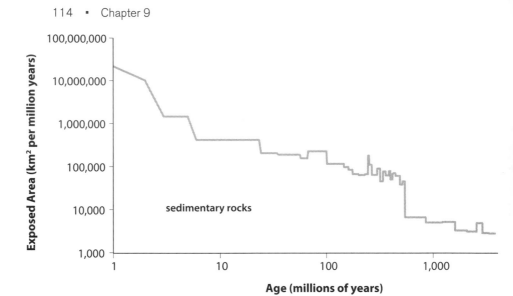

Figure 9.1. The area of exposed sedimentary rocks as a function of geologic age. The steep drop-off with time partly indicates the burial of rocks into Earth's crust; however, over long time scales, loss by weathering and erosion is probably the most important cause of the drop-off. From data kindly supplied by Bruce Wilkinson (see Wilkinson et al. 2009).

This is because through time, they have had a higher probability of being delivered to the zone of weathering at the Earth surface. Indeed, if you make a tally of all the rocks you can find on the Earth surface and group them into ages, you find that the abundance of rocks decreases as age increases (fig. 9.1). If you try to quantify this effect, about half of the rocks remaining from any time in Earth history will disappear through weathering (though some are buried in the crust) over a time scale of about 200 to 300 million years. This is approximately the same length of time as the Lomagundi isotope event, and I think it makes sense that the duration of the event could have been limited by the availability of old Archean rocks and the availability of recycled organic matter and nutrients. To be sure, this is all just speculation, but I would bet that the Lomagundi event will be the focus of intense scrutiny over the next few years, and I will be surprised if Holland's initial instinct, that it is related in some way to the GOE, is proven incorrect.

So, the GOE appears to have been followed by a massive burial of organic carbon as expressed in the Lomagundi isotope event. Dick Holland and his former postdoc Andrey Bekker have argued that the Lomagundi event built on the already higher oxygen levels of the GOE to produce oxygen levels similar to those at present. There isn't much evidence yet to back this up, but if oxygen became so high, did things settle back down after the Lomagundi event? What indeed was normal at this time in Earth history?

Many years ago I began to ponder exactly this point. The Lomagundi isotope excursion hadn't really entered the discussion yet, but the GOE had. Prevailing wisdom, stemming mainly from the contributions of both Dick Holland and Preston Cloud, was that the GOE ushered in oxygen levels high enough to ventilate the bottom of the ocean with oxygen. Thus, the idea that oxygen rose to high levels during and after the GOE isn't new. This oxygenation, in turn, oxidized iron from the oceans. Thus, the GOE called an abrupt end to iron-rich conditions in the deep oceans and hence the deposition of banded iron formations. High post-GOE oxygen levels could, therefore, explain an important geological observation; that is, the near cessation of BIF deposition following the GOE (fig. 9.2). Indeed, this view was presented in the last chapter.

While this idea of high post-GOE oxygen levels seemed reasonable enough, there were some parts of the idea that didn't sit well with me. For one, I had recently used a simple model of ocean chemistry, first introduced by Jorge Sarmiento at Princeton University, to explore how deep-water oxygen levels in the ocean might respond to changes in atmospheric oxygen concentrations. The model is very simple and only approximates the workings of the real ocean, but the result was pretty shocking. In order to fully oxygenate the oceans after the GOE, as Holland and others had suggested, atmospheric oxygen levels would need to rise to some 40% to 50% of today's value. Maybe Dick would have accepted this conclusion easily, but I couldn't. For one, as we will explore in the next chapter, there was accumulating evidence for a late Precambrian (some 600 million years ago) rise in atmospheric oxygen levels. This would be difficult to accommodate if oxygen already rose to near modern levels at the GOE. There is also a weaker argument explored in the next two chapters.[5] This argument is based on the idea that the widespread evolution of large(ish) mobile animals, much later in geologic history, occurred as oxygen levels rose to accommodate their rather high energetic demands. This was would be equally inconsistent

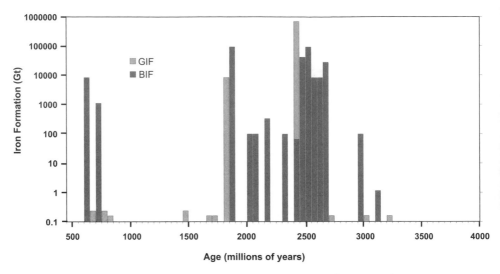

Figure 9.2. Age distribution (in gigatons, Gt, where one Gt is 10^9 metric tons) of banded iron formations (BIFs) and granular iron formations (GIFs). GIFs are more sand-like deposits and lack the banding of BIFs. Figure redrawn from Bekker et al. (2010), and modified from Raiswell and Canfield (2012).

with an earlier rise of oxygen to present levels. Finally, there was perhaps another way to explain the loss of the BIFs.

I had been compiling a history of sulfur isotopes through time while working as a research scientist at the Max Planck Institute for Marine Microbiology in Bremen, Germany. My real job was to explore the fate of organic matter in sediments and the various microbial processes responsible for metabolizing it, but the sulfur isotope history was engaging, and you might call it my night job. What emerged from this compilation was a picture of how the sulfur cycle had evolved through Earth history, and one striking feature was a big increase in the isotope difference between sulfate and sulfide around the GOE (recall sulfur isotopes from chapter 7). In fact, Eion Cameron from the Geological Survey of Canada, was the first to recognize this increase, but the new compilation emphasized its singularity (fig. 9.3). Eion had suggested that the jump was due to an increase in the sulfate concentration of the ocean in response to the GOE and the enhanced oxidation of sulfide minerals to sulfate during weathering on land. This is very consistent with what we discussed in the last chapter. The logic here is that sulfate-reducing bacteria preferentially reduce the light sulfur isotope in sulfate,

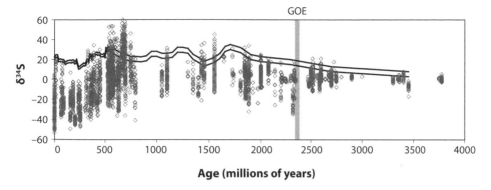

Figure 9.3. Compilation of sulfur isotopes through time. Diamonds represent individual analyses of sediment sulfides whereas the parallel lines reflect an estimate for the isotopic composition of seawater sulfate. The GOE is also marked. Figure modified from Raiswell and Canfield (2012).

^{32}S, compared to the heavier isotope, ^{34}S, as explored in chapter 7, but the extent to which they do so is greatly reduced at very low sulfate concentrations.[6] With my compilation in hand, I could therefore only agree with Eion, but one could take the argument another step. An increased flux of sulfate to the oceans should stimulate more sulfate reduction, producing more sulfide in the ocean. As a result, more dissolved iron would be removed in the formation of pyrite (FeS_2).

Thus, piecing the whole thing together, I reasoned that the GOE likely increased the flux of sulfate to the oceans through the oxidative weathering of sulfides on land, thus enhancing rates of sulfate reduction to hydrogen sulfide in the ocean. The GOE did not, however, at least in my thinking, produce enough oxygen to oxygenate the oceans. Therefore, dissolved iron was removed from the oceans not by reaction with oxygen as Holland and Cloud supposed, but rather by reaction with sulfide. So in my proposal, the deep oceans remained anoxic and became rich in sulfide, instead of becoming well oxygenated. Andy Knoll from Harvard University soon named this sulfide-rich oceanographic state the "Canfield Ocean," and for better or worse the name has stuck. At the time I made this proposal, however, there was not a stitch of evidence to back it up, so it became a high priority to look for evidence to either support or falsify the Canfield Ocean model and to more generally define the state of Earth surface and ocean chemistry after the GOE.

At about this time, Simon Poulton joined my lab as a postdoc. Simon worked as a PhD student with my old friend Rob Raiswell from the University of Leeds,[7] and like Rob, Simon knows his way around a pub. Also like Rob, he is a creative and resourceful scientist. So, to explore this Canfield Ocean problem, Simon and I focused on the last episode of major BIF deposition before its resurgence much later in the Neoproterozoic Era, which we will discuss in the next chapter. Our focus, therefore, became an episode of major BIF deposition between about 2.1 and 1.9 billion years ago as observed in several places around the world, but with particularly good representation in northern Minnesota and southern Ontario. This became our target.

Wait a minute, you might say, no matter what the model, I thought the GOE was supposed to have ushered in the end of BIFs, and here you're talking about BIFs some 300 to 400 million years AFTER the GOE? What gives? Good question, and before I can get to our work on the Canfield Ocean model, we must contend with these BIFs.

Indeed, these BIFs are quite telling. For one, deposition began either just after or near the end of the Lomagundi isotope excursion. Also, at least some of them deposited in very shallow water. Imagine taking a swim at the beach and coming home covered in rust! This is a critical point. You can imagine the scenario like this: dissolved iron, accumulated in the deep oxygen-free ocean, was transported up onto the continental shelf by a process called upwelling and was then carried across the shelf into shallow water.[8] This sounds simple enough, but if you recall from chapter 7, Fe^{2+} reacts readily with oxygen to form rust. Therefore, if oxygen concentrations were high like today, then the Fe^{2+} simply could not be transported from the deep ocean, across the continental shelf, and further to the beach. The only logical way I can think of for Fe^{2+} to persist over extended transport in the surface ocean is if oxygen levels were very low, perhaps only 0.1% of present-day levels or maybe even lower.[9] Indeed, some recent geochemical evidence provides independent evidence for low oxygen at this time. This evidence comes from the behavior of chromium isotopes. I won't go into it here, but you can learn more by following this endnote.[10]

Let's review. There was a substantial rise in oxygen defining the GOE. This may, in turn have led to the Lomagundi isotope excursion, which as we argued, was associated with high rates of organic matter burial and perhaps even higher concentrations of oxygen. This excursion was followed soon after by a crash in oxygen to very low levels and a return to BIF deposition. One can surmise that these low levels of oxygen were, in fact, a consequence of the Lomagundi excursion. It's easy to imagine that when the massive amounts of organic carbon buried during the excursion were brought into the weathering environment,[11] they would have represented a huge oxygen sink, drawing down levels of atmospheric oxygen. We appear, then, to have a veritable seesaw in oxygen concentrations, apparently triggered initially by the GOE.

Now back to Simon. If you recall, we were interested in any possible evidence to support or refute the idea that with an increase in atmospheric oxygen levels, sulfide expanded into marine waters, bringing an end to BIF formation. As mentioned, rocks in northern Minnesota and southern Ontario became our focus, and in particular, we looked toward the Gunflint Iron Formation deposited about 1.9 billion years ago. As just argued above, this BIF, and other BIFs from the same time, appear to have deposited during a period of very low atmospheric oxygen concentrations. But, what happened just after the Gunflint stopped depositing? To tackle this problem, we contacted the effervescent Phil Fralick from Lakehead University in Thunder Bay, Ontario. Phil probably knows the Gunflint Iron Formation better than anyone. During a memorable trip, Phil took us throughout the region to look at both the Gunflint and the sediments depositing just after. What struck us in the field was that the Gunflint gave way to a very black shale known as the Rove Formation. This was encouraging because black shales were often deposited in sulfidic waters.[12]

The rocks we saw in the field were great for understanding the relationship between the Gunflint and the Rove Formations, but not great for doing chemistry. As we have learned, such rocks lay exposed to the weather and rain, becoming oxidized in the process. Any pyrite, for example, originally in the rocks may well have been oxidized away. Luckily, Phil knew of fresh rocks, recently cored from underground, which nicely covered the transition from the Gunflint Iron Formation to the overlying Rove Formation. The core, it turns out, came from a curious farmer with a drill rig who once or twice a year drills a core through the rocks underlying his land in order to prospect for precious metals. When the cores turn up nothing of interest, he donates them to the Ontario

Ministry of Northern Development and Mines. The farmer's disappointment was our good fortune, and after carefully sampling the core, we returned to the lab for a suite of chemical analyses.

I won't dwell on the details of the analyses, but we subjected the samples to various chemical treatments.[13] Our analyses, in the end, told us that the Fe^{2+}-enriched ocean waters precipitating the Gunflint BIF gave way to sulfidic waters from which the Rove Formation deposited. This was very consistent with the Canfield Ocean model. Indeed, at about the same time we presented this, Gail Arnold was working with Ariel Anbar from Arizona State University and Tim Lyons from the University of California Riverside (and an old graduate student pal from Yale) and presented further support for the Canfield Ocean model from the isotopes of the molybdenum preserved in 1.6 billion year old black shales. For an explanation as to how this isotope system works, follow this endnote.[14]

The story seemed to be taking shape, but one should never get too comfortable with an idea. In a stroke of good luck, Phil Fralik identified a series of cores that represented environments ranging from the shallow waters where the Gunflint-Rove Formations deposited, to far offshore, where the waters were well over 100 meters deep. There was a total of over 100 kilometers between the cores. Through painstaking work, Phil was able to draw time lines between all of these cores, and Simon spent nearly a year working up all of the geochemical data, but it was well worth the trouble. What emerged was a spectacular, one of a kind picture of ocean chemistry stretching from nearshore to well out to sea (fig. 9.4). As we saw before, the Gunflint Formation gave way to the sulfidic Rove nearshore. Moving offshore, however, we saw that anoxia persisted, but we also saw that the sulfidic waters disappeared. Indeed, we saw that the sulfidic waters gave way to Fe^{2+}-containing waters, something like (but not completely like, as explored below) the waters from which the Gunflint BIF deposited. We call such Fe^{2+}-enriched waters "ferruginous." This is an amazing two-dimensional look at ocean chemistry deep in time. Hats off to Simon and Phil!

Others have also contributed to this story including Andy Knoll's group, Andrey Bekker and his colleagues, as well as Tim Lyons and his group. Altogether, we can paint a picture of, yes, more widespread marine sulfidic conditions compared to time before the GOE, but with lots

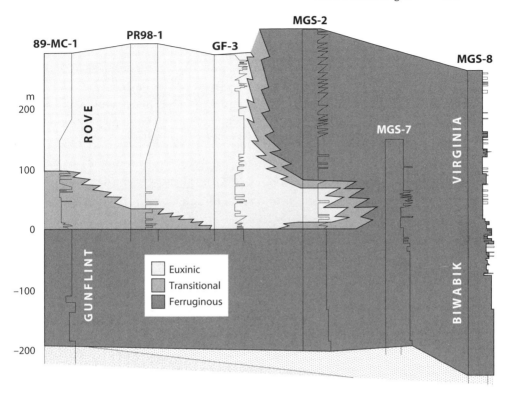

Figure 9.4. Development of ocean chemistry through the transition from the Gunflint Iron Formation (the Biwabik offshore) to the overlying Rove Formation (the Virginia Formation offshore). Time moves upwards in the plot (younger sediments on top) and we move further offshore in going from left to right. Figure reproduced from Poulton et al. (2010).

of evidence for deep-water ferruginous conditions as well. And ferruginous conditions can be found all the way until the time when animals first evolved some 1.3 billion years later. We'll look at when animals make an appearance in the next chapter, but let's reflect on where we stand relative to the original Canfield Ocean model. After the GOE, the deep ocean appears to have remained largely anoxic as the model predicted, and sulfidic conditions were more extensive as the model also suggested. However, these sulfidic conditions were not particularly extensive. Instead, ferruginous conditions seemed to dominate in the anoxic ocean.

If ferruginous conditions dominated, why then are BIFs exceedingly rare during this time in Earth history, as we saw graphically in figure 9.2? Good question. I believe that the original Canfield Ocean model still holds the answer. As the model suggested, increased oxygenation led to more pyrite weathering and a higher flux of sulfate to the ocean. This led to more sulfide production from sulfate reduction, expanded sulfidic conditions, and enhanced removal of Fe^{2+} from solution. Even so, overall, the flux of Fe^{2+} to the ocean was apparently still greater than the rates of sulfide production by sulfate reduction.[15] Thus, while sulfide could accumulate in those regions of the ocean where rates of sulfate reduction were high, in other regions, Fe^{2+} dominated, generating ferruginous waters. It appears, however, that there was enough sulfide produced by sulfate reduction to react with a sizable proportion of the dissolved Fe^{2+} from the oceans. This meant that while ferruginous conditions could still persist in large volumes of the anoxic ocean, the concentrations of Fe^{2+} were significantly lower than during times when BIFs were actively forming; quite simply, with lower Fe^{2+} concentrations, there was not enough iron dissolved in the ocean to support active BIF deposition. Thus, it appears that the Canfield Ocean model had it right in general, while not in all the specifics.

Perhaps most important to the present discussion, however, is that deep anoxic ocean waters appear to have been the normal state through Earth's middle ages. This implies that atmospheric oxygen concentrations were considerably lower than today. But how much lower? The original ocean modeling that started me on this track suggested that vast expanses of deep anoxic ocean water would develop as atmospheric oxygen dipped to some 40% to 50% of present levels. Unfortunately, the geologic record does not provide a more precise estimate. However, I believe that oxygen levels were much lower than this, probably less than 10% of present levels. This assertion is based not so much on the evidence we've discussed so far, but rather, on what comes in the next two chapters.

CHAPTER 10
Neoproterozoic Oxygen and The Rise of Animals

The Avalon Peninsula of southeastern Newfoundland is a remarkable place. Carved from glacial ice, its rugged terrain was further sculpted by wind, rain, and the sea. On the Avalon Peninsula you never feel far from the sea. It's also a place of tradition, old tradition. Already in 1534, the noted French navigator Jacques Cartier commented that the fish were so thick off the Newfoundland coast that "they slowed our ships in the water." He was not the first, however, to note the abundance of fish, and fishing settlements were already established on the Avalon Peninsula in the early 1500s. A colony followed in 1610 in Conception Bay (Cupar's Cove) in the northern part of the peninsula, and in 1621 another colony was formed at Ferryland in the southern part of the peninsula (fig. 10.1). The latter, which is presently under archeological excavation, was established by Sir George Calvert, later Lord Baltimore, and became the first permanent colonial settlement in Newfoundland. Indeed, Sir George moved to the settlement in 1627 but left the following summer, explaining to King Charles I:

> I have found by too dear bought experience, which other men for their private interests always concealed from me, from the middest of October to the middest of May, there is a sad face of winter upon this land, both sea and land so frozen for the greatest part of time as they are not penetrable, no plant or vegetable thing

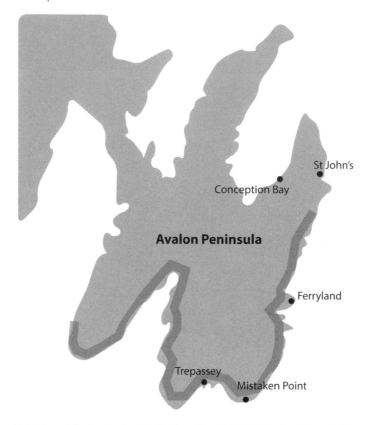

Figure 10.1. Map of Avalon Peninsula, Newfoundland, highlighting locations of places discussed in the text. The heavy gray line marks the Irish Shore.

appear out of the Earth until it be about the beginning of May, nor fish in the sea, besides the air so intolerable cold as it is hardly to be endured.[1]

Much later, from the mid-eighteenth to the mid-nineteenth centuries, the Irish, perhaps more attuned to harsh conditions, immigrated en masse to Newfoundland and especially to the Avalon Peninsula. In 1836, a government survey revealed that half of the total population of Newfoundland was of Irish descent, with most of them living around St John's. One can still see the Irish influence today, especially in southeastern part of the peninsula, the so-called Irish Shore. Here, Irish music is enjoyed and the residents speak with an Irish accent with such character that in many cases it can still be traced to the original Irish county where their ancestors lived nearly two hundred years ago.

It is here, in the small town of Trepassey, on the southern part of the peninsula, that Simon Poulton and I warm our frozen bodies with hot cups of coffee and a delicious chicken soup made by Paula Carew, a short, bubbly dynamo of a woman and proprietor of "First Venture," a Trepassey eating tradition. We were too long and too late in the field today. Paula floats over to our table and asks in a rich Irish brogue if we would like to follow our coffee with something stronger, perhaps a beer or maybe a whiskey? We nod emphatically to both. We are part of a field trip organized by Guy Narbonne of Queens College, Ontario, and Jim Gehling from the South Australian Museum in Adelaide. Guy and Jim are both world experts on a fascinating group of marine organisms that first appear in the fossil record around 580 million years ago. El-kanah Billings, a prominent Canadian paleontologist, was the first to describe a member of this group in 1871. In fact, he found a fossil type called *Aspidella terranovica* in open view in black shales outcropping onto Duckworth Street, in downtown St. John's. You can still see them there today. Billing also noted that these fossils were from sedimentary strata deposited before the appearance of abundant animals in the Cumbrian Period, a topic we will come to later.

Fossils from this time period came to prominence much later, however, when Reg Sprigg in 1946 stumbled across some strange fossil forms he interpreted as ancient jellyfish-like animals from the Ediacaran Hills of the Flinders Ranges in South Australia. These late-Precambrian fossils, and many others of similar age, have become collectively known as the Ediacaran Fauna. Taken together, these fauna represent a wide range of very different types of organisms, but they also share some common attributes (fig. 10.2). They are mostly in the centimeter to tens of centimeter size range, lived on the seafloor either lying down or standing on some kind of a stalk, were mostly immobile, exhibited complex but semiregular morphologies, and with a few exceptions, cannot be easily assigned to any modern or ancient animal group. Therefore, Sprigg's first impression of ancient fossil jellyfish was probably incorrect, leaving the question open as to what indeed the Ediacaran Fauna represent.

Figure 10.2. Various Ediacaran fossils from Mistaken Point, Newfoundland. A) *Charniodiscus*, B) spindle-shaped rangeomorph, C) another type of spindle-shaped rangeomorph. Photos from the author.

Needless to say, there is active scientific debate on this issue. Despite their unusual forms, until the late 1980s, the Ediacaran Fauna were usually thought to represent ancient, primitive animal forms. Debate was sparked when leading paleontologist Dolf Seilacher from Tubingen, Germany, reinterpreted these fossils as something completely different. He argued that, instead of animals, they were long extinct varieties of living organisms, a result of failed lineages with no successors. In Dolf's view, they had nothing to do with animals at all. Dolf has a special eye for how animals, long extinct, dug, crawled, and fed in the muds of the ancient world, and what Dolf said was therefore taken very seriously. Seriously, yes, but not all, and perhaps not even most, have been convinced by his reinterpretations.

Indeed, recent thinking has returned to animal interpretations for most of these organisms, albeit not animal groups that are necessarily represented today. For example, let's take a weird one. An Ediacaran faunal type called *Charniodiscus* is found on the Avalon Peninsula among the very rocks that Simon and I went to explore (fig. 10.2). In life, it was 10 or more centimeters in length and was firmly rooted on the seafloor. It was constructed of a holdfast and stem connected to a frond that gently swayed in the ocean currents, presumably collecting small particles or dissolved organic material from the water. The water was quite deep where these organisms lived, so they lived well out of reach of any sunlight. In summary, these organisms were large, they exhibited complex morphology constructed from many different cell types, as animals do, and they fed like animals. According to Andy Knoll, it is difficult to imagine how such organisms could have grown without an epithelial covering of cells,[2] another attribute of animals. Finally, because of the deep, light-free water depths where they grew, any plant or algal affiliation is out of the question. So there you have it, many animal-like attributes, but not much like any animal you or I might know. And so it goes with many of the Ediacaran Fauna.[3]

So, what brought Simon and me to the Avalon Peninsula? As it turns out, rocks on the Avalon Peninsula house the oldest known representatives of the Ediacaran Fauna. These so-called rangeomorphs (fig. 10.2) date back to 575 million ago and appear relatively soon after the end of the Gaskiers glaciation some 580 million years ago.[4] Was this a coincidence? Could we identify any geochemical triggers that might correlate with or even help to explain the rather sudden appearance of animal-like forms? Our focus was on oxygen. There has been a long prejudice, starting with a paper published in 1959 by J. Ralph Nursall from the University of Alberta, that the evolution of animals, at least that of large forms with relatively high oxygen demand, was spurred by an increase in the oxygen content of the atmosphere. One can discuss long and hard the relationship between animal evolution and oxygen, and we'll entertain some of that discussion below, but Simon and I reasoned that looking for signs of changes in ocean chemistry would be a good place to start. These rocks on Avalon Peninsula seemed like a good place to begin.

Rather remarkably, throughout the southern part of the Avalon Peninsula, one can sample rocks more or less continuously from a time beginning before the Gaskiers glaciation, through the Gaskiers, and extending to some 20 million years later. Importantly, one can sample with constant reference to the emergence of Ediacaran life, because fossils are abundant and well preserved throughout these rocks. Furthermore, Guy and Jim are the perfect trip leaders. They've worked on these ancient sediments for years and probably know them better than anyone. They know the fossils, where to find them, and where on the seafloor these organisms lived. They led many lively discussions on how the organisms fed, their possible animal affinities, and how they were preserved. We approached the famous Mistaken Point location at sunset. This is the best-known Ediacaran fossil site on the Avalon Peninsula. Guy likes to arrive here at sunset because in many cases the fossils here, and throughout the peninsula, are only faint impressions. Difficult to see at high noon, they emerge in all their glory as they cast small shadows in the light of the setting sun.

Simon and I collected hundreds of samples.[5] Back in the lab, we conducted the same type of chemical analyses, the so-called Fe speciation, which I described in the last chapter for the transition between the Gunflint BIF and the overlying Rove Formation. We found that before and during the Gaskiers glaciation, the deep waters of the ocean in this part of the world were anoxic and ferruginous.[6] Indeed, you can see this just by looking at the glacial deposits of the Gaskiers, which are blood red with iron (plate 9). Subsequent work by Simon, myself, and many others showed that these ferruginous deep water conditions were com-

mon, dating back to at least 850 million years ago.[7] Indeed, these ferruginous waters may also link with the common occurrence of anoxic
and often ferruginous conditions described in the last chapter during
the middle parts of the Proterozoic Eon. However, and critically, Simon
and I, together with Guy, also found that just after the Gaskiers glaciation, the deep ocean waters of the Avalon Peninsula became well
oxygenated. This is, as far as I know, the first evidence for oxygenation
of the deep ocean during Earth's long history.[8] And this may not have
been just a local event. Further evidence for deep water oxygenation at
about this same time was found in rocks from western Canada by Yanan
Shen, working with Andy Knoll and Paul Hoffman from Harvard, and
also by our group.

Therefore, the Ediacaran Fauna of the Avalon Peninsula emerged
into an ocean undergoing oxygenation. Indeed, other lines of evidence
seem to support such a conclusion. For example, Tim Lyons's group
has very recently found that concentrations of molybdenum in black
shales increased dramatically around 630 million years ago in the late
Neoproterozoic Era, somewhat predating our evidence for deep ocean
oxygenation (fig. 10.3). This increase in molybdenum abundance implies
higher seawater molybdenum concentrations, which in turn implies less
efficient removal of molybdenum from seawater. The most efficient removal pathway for molybdenum occurs under anoxic conditions, particularly during interaction with dissolved sulfide. Piecing this together,
an increase in molybdenum abundance in black shales implies less anoxic removal of molybdenum in the oceans. This, in turn, can be reasonably tied to an increase in ocean oxygenation. The same is true for
molybdenum isotopes (see endnote 14, chapter 9), which show a big
jump about this same time, as detected by my postdoc Tais Dahl and
then PhD student Emma Hammarlund (fig. 10.3). Tais and Emma also
interpret this jump as reflecting an increase in the removal of molybdenum under oxic conditions, and consequently an increase in ocean
oxygenation.

Different geochemical indicators provide somewhat different answers,
but an increase in ocean oxygenation seems apparent in the late Neoproterozoic Era, beginning perhaps as early as 630 million years ago and
expressing itself in local indicators of oxygenation by 580 million years
ago. All of this is timed roughly with the emergence of macroscopic

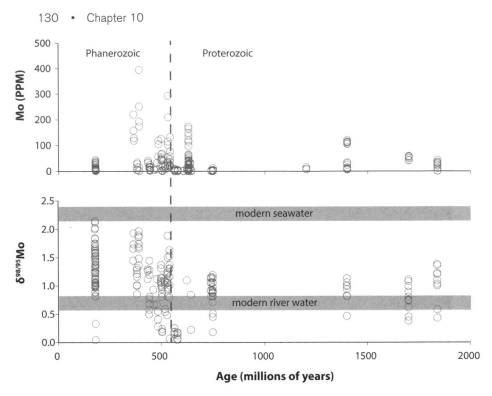

Figure 10.3. Concentrations and isotopic compositions of molybdenum over the last two billion years. Data from a variety of sources through a compilation kindly made available by Tais Dahl with additional data from Sahoo et al (2012).

animals. One possible way to explain the emergence of larger animals would be an increase in atmospheric oxygen concentrations. We will therefore follow this line of reasoning and try to deal with three related issues: (1) If oxygen rose, how high did it go? (2) What caused the rise? (3) What might be the relationship between rising oxygen and animal evolution?

Let's begin by taking on the first issue. Independent of any considerations about animals and their needs for oxygen, can we say anything about late Neoproterozoic atmospheric oxygen levels? Unfortunately, our geochemical tool kit is not yet developed enough to provide a specific answer, but with some careful reasoning, we can maybe come up with some rough estimates. Let's start with the molybdenum isotopes. While it's true that molybdenum isotopes point to an increase in the removal of molybdenum from the oceans into oxic marine settings in the late Neoproterozoic Era, the isotopic composition of molybdenum remains much lower than we find today. This suggests that around the time animals emerged, marine oxic conditions were not as widespread as today. This would in turn imply lower oxygen levels, maybe even much lower, than we have at present.

Not too low though. We also attempted to estimate levels of atmospheric oxygen when we first described evidence for deep-water oxygenation from the rocks of the Avalon Peninsula, as explored above. The argument went like this. In the modern ocean there is typically a zone of minimum oxygen concentration developed in a depth range from a few hundred meters to about 1500 meters. This is approximately the depth range where the organisms represented by the Avalon fossils lived. The oxygen minimum results from the convergence of two factors. The first is the settling of organisms from the surface ocean; these consume oxygen as they decompose. The second is the sinking of cold, oxygen-rich surface water from the polar regions of the oceans into the ocean depths. This water forms the deep water of the oceans. Very little organic matter survives the long transit from the ocean surface to these depths, so little organic matter decomposition takes place, allowing oxygen to persist. Consequently, in the modern ocean anyway, we have high oxygen levels in the ocean surface, high levels at great depth, and lower oxygen levels in between, which generates an oxygen minimum zone. The magnitude of this oxygen minimum varies from place to place, but in the modern world, the decrease in oxygen concentration is, at a minimum, around 40 micromoles of oxygen per liter. In today's oceans, the deep waters start with an oxygen concentration of about 325 micromoles per liter,[9] and a 40 micromole per liter decrease in concentration represents 12% of this modern deep water value. We assumed then that the oxygen decrease where the organisms represented by the Avalon fossils lived was at a minimum about 40 micromoles per liter (possibly more). Finally, to give the organisms a bit to live on, we argued that some oxygen should be left after this 40 micromoles per liter was used. In the end, we estimated that the organisms lived in a world where atmospheric oxygen levels were 15% or more of present levels.[10] Less than this, and a lack of oxygen would have suffocated the organisms at the depth they lived, but higher oxygen levels would have still enabled life. So, if we were right, oxygen was something like 15% or more of modern values during the late Neoproterozoic Era, but still at levels considerably less than those of today. This is about the best I can offer at present with the evidence at hand, but in the next chapter we may be able to check this estimate with other approaches.

If there was a rise in atmospheric oxygen levels during the late Neoproterozoic Era, there must have been a cause. Some years ago, the answer seemed clear. In a major geochemical breakthrough, Andy Knoll, in 1986, provided the first detailed carbonate carbon isotope results from Neoproterozoic rocks. These rocks displayed a preponderance of ^{13}C-enriched values, and if you remember from chapter 7, this means high rates of organic carbon burial. That also means high rates of oxygen release to the atmosphere. Supercontinent Rodinia was undergoing breakup into smaller continental units during the late Neoproterozoic, and Andy reasoned that with the ensuing increase in coastal area,[11] there would be more sediment deposition, giving higher rates of organic carbon burial. Refining this argument, Lou Derry from Cornell, but a Harvard postdoc at the time, introduced other isotope systems that helped to isolate the peak of organic carbon burial to around 580 million years ago.[12] This was perfect. Problem solved.

But as often happens in science, an interesting idea is subject to testing and the accumulation of more data, and in this case, lots more data. Whereas Lou had perhaps one hundred carbon isotope measurements to consider for his modeling, there now exist thousands (fig. 10.4). These new data have preserved many of the broad features of the carbon isotope record that Lou modeled, but the dating has changed and new features have emerged that considerably confound the picture. One of these is a big excursion to very ^{13}C-depleted values known as the Shuram-Wonoka carbon isotope anomaly (fig. 10.4). The dating on this anomaly and its duration are uncertain, but most agree that it comes sometime after the Gaskiers glaciation at 580 million years ago, and before 551 million years ago. It has now been identified in many places from around the world, and it seems to be a robust feature of the late Neoproterozoic carbon cycle.[13]

The problem with this anomaly is that it's almost impossible to understand how it originates given our normal picture of how the carbon cycle works. The issue is where to get all of the ^{13}C-depleted carbonate that winds up in the limestone. There are a variety of possible solutions

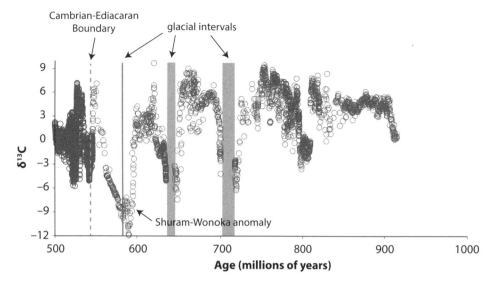

Figure 10.4. Neoproterozoic record of inorganic carbon isotope distribution. The Shuram-Wonaka anomaly is indicated as well as the major episodes of glaciation. The relationship of the Shuram-Wonaka anomaly to glaciation (it could well come after) as well as the length of the anomaly are highly uncertain. Data compiled and kindly made available by Galen Halverson of McGill University and Matt Saltzman from Ohio State University.

to this problem, including an ingenious proposition by Dan Rothman from MIT. He imagined that at this time in Earth history, the ocean was much like an organic soup, containing huge concentrations of dissolved organic matter. Dissolved organics should have the same ^{13}C-depleted isotopic signal of algae as we explored in previous chapters. Dan argued that the occasional oxidation of this organic soup produced a huge amount of ^{13}C-depleted CO_2. Indeed, Dan argued that enough ^{13}C-CO_2 could be produced to give us the Shuram-Wonaka anomaly. This is a brilliant idea, but I've never quite understood how this oxidation episode was started. For this and other reasons, Christian Bjerrum and I have proposed another solution involving the oxidation of a huge methane pool (containing even more ^{13}C-depleted carbon than dissolved organics), which you can read more about if you follow this endnote.[14] In any event, understanding the Shuram-Wonaka anomaly requires some lateral thinking, but these solutions all come with a cost, at least in terms of oxygen. So, whether we're talking about the oxidation of dissolved organics or methane to form the Shuram-Wonaka ^{13}C

anomaly, a huge amount of oxygen is required to perform the oxidation,[15] producing a huge oxygen sink.

Let's see where we stand. Geochemical evidence suggests increased oxygenation of the oceans by 580 million years ago, and maybe somewhat earlier, and in approximate concert with the expansion of the Ediacaran Fauna. Perhaps, as Andy Knoll and Lou Derry suggested, this was driven by high rates of organic matter burial associated with the breakup of supercontinent Rodinia. Sometime soon after this, however, the Shuram-Wonoka anomaly suggests a huge oxygen sink and the probability of significant oxygen drawdown in the atmosphere and oceans. We don't know how far oxygen might have decreased through this anomaly, but apparently not so low as to affect the respiration of early animals.

What happened next? Did oxygen rise again to pre-Shuram-Wonoka levels? Some of the geochemical evidence as outlined above might suggest this, but the carbon isotope record isn't revealing much. In fact, Lou's model only showed a transient increase in oxygen liberation around 580 million years ago. After this, organic carbon burial and oxygen liberation reverted to background levels. Also, to confuse things even further, the carbon isotope record reveals another big ^{13}C depletion (fig. 10.4) just at the Cambrian-Precambrian boundary, in concert with the great expansion of animal life. Taken at face value, animal life increased in magnitude and diversity as oxygen release to the atmosphere waned. There is much here we don't understand.

Well, what about animals? As mentioned earlier in this chapter, there has been a prejudice dating back to Nursall that animal evolution, at least macroscopic motile animals, was enabled by an increase in atmospheric oxygen concentrations. Indeed, in 1982 Bruce Runnegar of UCLA tried to put some numbers on this. He took the Ediacaran fossil *Dickensonia* as his example, which he viewed at the time as an example of an ancient annelid worm. Most would probably disagree with this view now, but for the purpose of the calculation it really doesn't matter very much what *Dickensonia* was. What is important is the reasonable assumption made by Bruce that *Dickensonia* obtained its oxygen by passive diffusion across its outer surface. With this assumption, Bruce concluded that at least 10% of present-day oxygen levels would be required to maintain *Dickensonia*'s respiration. Ignoring the possible oxygen gymnastics associated with the Shuram-Wonoka anomaly, the oxygen demand for *Dickensonia* seems to fit with our geochemical arguments suggesting that around 580 million years ago oxygen rose to at least 15% of present levels. Sounds pretty good. In this scenario, the evolution of large and possibly motile animals like *Dickensonia* was enabled by a rise in oxygen to levels conducive for their livelihood. In the words of Andy Knoll, motile animals evolved into a "permissive environment."

This type of logic makes some paleontologists cringe, and particularly Nick Butterfield from Cambridge University. I can almost see him shudder every time he reads about a possible relationship between oxygen concentrations and animal evolution, especially if the argument comes from a geochemist. Nick's view is that oxygen should be left out of it. He argues that animals themselves have engineered the conditions of environmental change, and any evolutionary developments that we may attribute to a "permissive environment," were made permissive by animals themselves. Nick wrote the following in a 2011 article:

> By facilitating and forcing the diversification of, for example, eukaryotic phytoplankton, large body size, bioturbation and biomineralization, early animals reinvented the chemical interchange between the biosphere and the planet. In this light, the biogeochemical perturbations of the Ediacaran-Cambrian interval are more likely to be top-down consequences of animal evolution than its bottom-up cause.

Nick may be surprised to hear that I like this idea very much. There can be no question that animals themselves (including the obvious recent impact of humans) have fundamentally influenced the biogeochemical cycling of elements and the chemistry of the oceans. In this view, the oxygenation we see at about 580 million years ago, or perhaps a bit earlier, might be better interpreted as a redistribution of oxygen in the oceans due to animal activity, rather than an increase in the levels of atmospheric oxygen. Indeed, this idea dates back to Graham Logan, John Hayes, Glenn Hieshima, and Roger Summons; they argued in 1995 that the evolution of tiny animal plankton, so-called zooplankton, completely changed the carbon cycle and the distribution of oxygen in the oceans. The idea is that zooplankton produce fast-sinking fecal pellets. These would decompose less in the upper layers of the ocean as they sink to the sea bottom when compared to the smaller, slowly settling microbial biomass, which previously dominated the oceans.

You can think of it like this. Imagine you have a bucket with a small hole near the bottom letting water out. If you stood on a high balcony and you lowered the bucket slowly, you'd lose lots of water as it approached the ground, and if you lowered slowly enough, you might lose all the water while the bucket was still in the air. But, lower quickly and much less water will be lost from the bucket on its trip to ground. Now, the water loss from our bucket is analogous to the oxygen used during the decomposition of organic matter as it settles from the upper ocean to the ocean depths. If the organic matter settles quickly, less oxygen is used through its decomposition in the upper portions of the ocean (by analogy, less water is lost from the bucket) as compared to the situation where the organic matter settles slowly. In this second case, lots more oxygen will be used in the upper reaches of the ocean as the organic matter gradually settles and decomposes.

One could thus imagine that with an increase in the settling rate of organic matter, less oxygen would be used during its decomposition in the upper depths of the ocean (meaning the upper several hundred meters). This would result in an increase in the oxygenation of these waters compared to when organic matter settled slowly. Indeed, an increase in the oxygenation of the oceans is what our chemical proxies tell us, but if we follow the logic outlined just above, such an increase could maybe occur without any change in atmospheric oxygen levels. If this was true, it might help to explain our difficulty in recognizing the driving forces for a late Neoproterozoic increase in atmospheric oxygen. Indeed, maybe there wasn't an increase.

Graham Logan and his colleagues also offered some support for their idea that animal evolution might have changed the carbon dynamics of the oceans. They looked at organic biomarkers and found very little evidence for the preservation of algae-produced organic matter from the upper water column photosynthetic organisms in sediments older than about 590 million years. However, in sediments younger than about 530 million years old, such biomarker remains were found. There is a big and important gap in the data between 590 and 530 million years ago, but Logan and colleagues nonetheless argued that before 590 million years ago photosynthetic organisms largely decomposed during their slow fall to the seafloor. In contrast, by 530 million years ago, zooplankton and other pelagic animals emerged into the marine ecosystem and generated rapidly settling fecal pellets; these transported the remains of surface-living phytoplankton to the seafloor, to be discovered by geochemists half a billion years later.

This is an attractive hypothesis, but it still needs testing. The history of zooplankton evolution is poor to nonexistent. Such small organisms leave a lousy fossil record. Nick Butterfield argues, however, that even if their fossil record is bad, one still might be able to probe for their presence by focusing on what they might have eaten. In this case, there is a rich record of potential foodstuffs, the so-called acritarchs, which are believed to be the preserved casings of ancient algae or their resting-stage equivalents. These undergo a dramatic increase in ornamentation some 630 million years ago, which in Nick's view, signals a defense against predation by newly evolved zooplankton. This timing wouldn't quite fit Logan's results. If you recall, he and his colleagues argued that the influence of animals on carbon cycling came later, sometime after 590 million years ago.

Clearly, our understanding of the relationship between oxygen and animals is fuzzy. Many details need to be worked out. Among other things, we need to explore in detail how a changing carbon cycle, driven by animal evolution, would influence ocean oxygenation and its record, which are revealed by many different proxy indicators. This will likely require ocean modeling. However, it seems a real possibility that the story of ocean oxygenation and its relationship to animal evolution has at least as much to do with what was happening in the ocean as what was happening in the atmosphere. And if this is true, it's equally possible that motile animals evolved into an environment that was already "permissible" for some time before their appearance.[16] It will be very exciting to see how this story plays out.

CHAPTER 11
Phanerozoic Oxygen

I applied to five graduate schools in all, and four of them were in the Northeast: Columbia University, Yale University, Woods Hole Oceanographic Institute, and the University of Rhode Island. I figured I could visit them all if I took about 10 days or so. So, I packed my VW bus with a sleeping bag, a few things to eat and drink, and set out on a road trip from my apartment in Oxford, Ohio. At Columbia, or more precisely the Lamont-Doherty Geological Observatory (now called Lamont-Doherty Earth Observatory), I was scheduled to meet with the eminent oceanographer Wally Broecker. Something, I can't quite remember what, took Wally out of town the day I arrived, but I had a terrific day with his good colleague Tara Takahashi and whole range of very impressive students and staff scientists. My next stop was Yale, where I arranged to meet with Bob Berner. After arriving at Kline Geology Labs, a soulless, windowless,[1] brick and concrete monstrosity designed by a world famous architect (the late Philip Johnson), I was led to Bob Berner's office. Here, a tall man in a blue turtleneck shirt and a broad smile leaped from his desk and greeted me with a handshake. Bob spent a great deal of time with me. He showed me the labs, took me to lunch, and explained in detail all of the projects going on in the lab. He introduced me to Rob Raiswell, a life changing experience (see endnote 7, chapter 9). He also explained that I had just missed Bob Garrels, who visited Yale for three months every year. Bob Garrels (who we met first in chapter 5), was one of my scientific heroes even back then, before I became immersed in geochemistry. A chat with graduate students Bernie Boudreau and Mike Velbel finished the day; they assured me that New Haven was a terrible place to live, and given the choice, I'd be crazy to come to Yale.

Before my search for a graduate school, I spent two years working with Bill Green of Miami University, Ohio, on the chemistry of Lake Vanda, a permanently stratified sulfidic lake in McMurdo Valley, Antarctica. My work included three months on the ice in Antarctica, and the rest of the time in the lab; I was otherwise carefree, on the loose, and having the time of my life. Through my work on Lake Vanda, I realized that I wanted to pursue the role of microbial processes in shaping the chemistry of the environment. Bob was a well-known expert in this area, and the projects he described excited me very much. He had also explained, with great animation, some new modeling work he was doing with Bob Garrels. He explained something about a quantitative accounting of the history of atmospheric CO_2 concentrations through time as controlled largely by geologic processes. I didn't quite get it then, but with Bob's enthusiasm it seemed important. Little did I know then that the two Bobs (Berner and Garrels), were opening up a whole new understanding of how the various components of the Earth system interact to regulate atmospheric and ocean chemistry. This was big thinking of the highest order.

A few months after my road trip, the letters started to roll in: rejection, University of Washington; acceptance, University of Rhode Island; rejection, Woods Hole; and finally, the last two, acceptance at both Yale and Columbia. This posed a dilemma. I was deeply impressed with Bob and his approach, but Wally Broecker was (and is) one of the most creative and influential climate researchers working, and the Lamont-Doherty Earth Observatory is one of the best oceanographic institutions in the world. The choice was tough, but my heart said Yale, so there I went.

My PhD evolved into a study of modern marine sediments, and in particular, the roles of both iron-driven and sulfur-driven microbial processes in controlling sediment chemistry and the preservation of organic carbon in sediments. This was a field and a laboratory study involving the collecting, slicing, dicing, and analysis of various marine muds. Bob, on the other hand, long out of the lab, was busy on a much grander scale trying to conceptualize how the Earth system worked. Not many can do this.[2] It requires the special skill of extracting the simple truths from complex behavior and figuring out how these truths link together quantitatively. You also need an incredible base of knowledge. To model atmospheric CO_2 concentrations, one must recognize that a combination of multiple processes must be considered. These include weathering on land, which is influenced by a variety of parameters, like temperature, sea level (as it controls continental area), the hydrologic cycle, continental elevation, and the type of terrestrial vegetation.[3] Atmospheric CO_2 levels are also controlled by rates of volcanic CO_2 input, which are in turn controlled by rates of heat loss from the mantle, the types of sediment being subducted back into the mantle, and the rates of subduction. And these aren't even all of the considerations. But Bob (together with Bob Garrels and also Tony Lasaga) figured out first, which processes of the Earth system should be considered as controls on atmospheric CO_2, and second, how to model them.

I was lucky enough to sit on the sidelines through this incredible intellectual exercise. I remember thinking, how brash, but exciting, to put actual numbers on the history of atmospheric evolution; this is a history that is long lost to direct measurement (wouldn't that time machine be handy!) but reproduced on a graph you can hold in your hands. I would have wanted to be part of it, but I lacked the background and skills. Bob was also insistent that these were not student projects. They were way too difficult and way too risky.

Near the end of my PhD, though, I did get a chance to be involved in a small way. At this point, Bob was trying to model the history of atmospheric oxygen. The oxygen modeling built on the intellectual framework established during the modeling of atmospheric CO_2, but with substantial contributions from other places as well. During this time there was a ferment of ideas as to how the carbon, sulfur, and other cycles might be linked and quantitatively understood. Indeed, at many of the international geological congresses, the main players in this field would hold "cycle rallies" in various hotel lounges, where they would share ideas fueled by ample quantities of pizza and beer. Bob Garrels was the father figure to this crowd, and Bob Berner's oxygen modeling drew heavily from contributions by Garrels, as well as Abe Lerman of

Northwestern University, Fred Mackenzie at the University of Hawaii, and Lee Kump, one of Bob Garrels's last PhD students and now at Penn State University.

In fact, before Bob Berner began modeling atmospheric oxygen, Lee Kump and Bob Garrels published an earlier modeling attempt, producing a history of atmospheric oxygen concentrations through the latest 100 million years of Earth history. This is a wonderful piece of work, where oxygen liberation rates to the atmosphere were quantified from the carbon and sulfur isotope record (as explored in chapters 8 and 9), and expressions were derived to quantify oxygen removal rates during the weathering of carbon and sulfur-bearing rocks. The oxygen concentration in the atmosphere became the kinetic balance between the rates of oxygen production and consumption. In general, Kump and Garrels found that oxygen concentrations were elevated when the isotope record indicated high rates of oxygen liberation to the atmosphere and lower when the isotope record indicated lower rates of liberation. This is perhaps not too surprising, but the challenge, as beautifully embraced by Kump and Garrels, was to figure out how all of these processes were quantitatively linked and how oxygen concentrations could be extracted from the isotope data.

Indeed, Bob's first entrance in oxygen modeling built solidly on this work but added an important extra component, which is the idea of rapid recycling. This idea was discussed in chapter 5, and the basic point is that the most recently deposited sediments will also be the most prone to weathering through processes like sea-level change or uplift of the land. Thus, through rapid recycling, high rates of oxygen production through the burial of organic-rich sediments will quickly lead to high rates of oxygen consumption through the exposure of these organic-rich sediments to weathering. From a modeling perspective, and as explored in chapter 5, rapid recycling (along with a number of negative feedbacks) helps to dampen oxygen changes. This is important because the fluxes of oxygen through the atmosphere during organic carbon and pyrite burial, and by weathering, are huge compared to the relatively small amounts of oxygen in the atmosphere. Thus, all of the oxygen in the present atmosphere is cycled through geologic processes of oxygen liberation (organic carbon and pyrite burial) and consumption (weathering) on a time scale of about 2 to 3 million years. This may seem long to you, but over geologic time scales this is short, and small imbalances between the liberation and consumption rates of oxygen are quickly channeled into rapid changes in atmospheric oxygen concentration.

In addition to rapid recycling, Bob Berner had another great idea. He reasoned that there may be more direct ways than the isotope records to obtain a history of carbon and sulfur burial. To see how this works, we start with Aleksandr Borisovich Ronov, a famous Russian geologist.[4] He spent much of his career carefully compiling geologic maps of a group of rocks known as the Russian platform and from other rocks around the world. Ronov used these maps to assess the volumes of the different types of rocks preserved through geologic time. Thus, Ronov determined, among other things, the volume of sediments preserved from marine environments, from continental environments, and the amount preserved from coal deposits. This data set (and much of the other data that Ronov compiled) is a veritable gold mine of information, and a graph showing the distribution of different rock types through the Phanerozoic Eon is shown in figure 11.1.

Bob Berner noted that these different types of sediment each carried, on average, different quantities of organic carbon and pyrite sulfur. For example, coal deposits are rich in organic carbon and generally poor in sulfur,[5] while other sediments deposited on the continents (frequently comprised of sands and gravel) are poor in both organic carbon and pyrite sulfur. Marine sediments have intermediate organic carbon levels and elevated pyrite sulfur contents compared to terrestrial deposits. All this means that if the average sulfur and organic carbon contents of these rock types are known, as well as their deposition rates through time, then rates of oxygen liberation to the atmosphere can be directly determined. Bob assumed that the relative distribution of the preserved rock types through time, as revealed by Ronov's compilations, reflected their relative distribution at the time they were deposited. Bob also used a variety of scenarios to calculate total rates of sediment deposition. These scenarios ranged from constant rates of total sediment deposition through time to variable rates,[6] but in the end, variability in total sediment deposition rate had little influence on the model outcome.[7]

Thus, with estimates of carbon and sulfur contents in the different sediment types, and with a constant rate of sediment deposition, Bob calculated rates of organic carbon and pyrite sulfur burial through time.

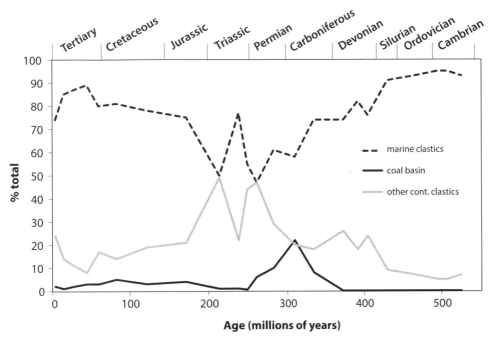

Figure 11.1. Distribution of sediment types through the Phanerozoic Eon. Data from A. B. Ronov as summarized in Berner and Canfield (1989).

We need one small additional detail. After some initial model runs, it seemed that some feedback in the sulfur cycle would be important. That was my job, and my modest contribution to the study. I figured out that an oxygen dependence on sulfur burial in marine sediments would be reasonable. This idea was explored in chapter 5, and the basic idea is that as atmospheric oxygen concentration lowers, there should be a spread of anoxic, sulfidic, conditions in the oceans. This leads to an increase in pyrite burial rates and hence an increase in rates of oxygen liberation to the atmosphere. The effect is a negative feedback; enhanced rates of sulfur burial under low oxygen conditions helps oxygen from getting too low.

To get some sense for the model results, rates of organic carbon burial calculated from the sediment abundance trends are shown in figure 11.2. These are also compared with organic carbon burial rates calculated from modeling carbon isotopes, as Bob did in his first study. To my eye anyway, there is a remarkable similarity between these two calculation

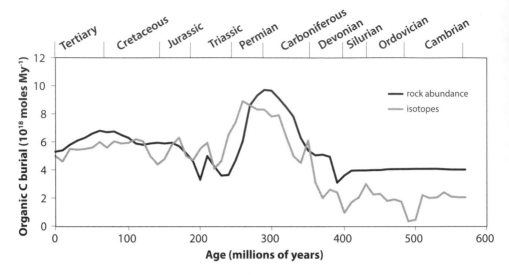

Figure 11.2. Rates of organic carbon burial calculated from both rock abundance data and from isotopes. Redrafted from Berner and Canfield (1989), using the time scale known at the time of the publication; most of the period boundaries have changed somewhat in time since then (see Preface).

results. This makes one think that, at least through the Phanerozoic Eon, both the geologic record of preserved sediment types and the carbon isotope record are telling us something similar and fundamental about the carbon cycle.[8] This puts us in a good position to perhaps say something important about the evolution of oxygen concentrations through the Phanerozoic Eon.

This also brings us to the model results, which are shown in figure 11.3. The gray area in the plot marks Bob's best estimate of the likely range in oxygen concentration as revealed by various sensitivity analyses, while the line is Bob's view of the best model result. The model clearly shows variations in atmospheric oxygen content. Sensitivity analyses revealed that rapid recycling was important to dampen fluctuations in atmospheric oxygen content and also that the organic matter content (in particular) and pyrite sulfur content (of lesser significance) of the various sediment types mattered a great deal. In contrast, the area of continent available for weathering mattered little, as did the total rate of sedimentation as explained above.

What mattered most, though, was the type of sediment deposited. The most obvious feature of the plot is the large positive oxygen excur-

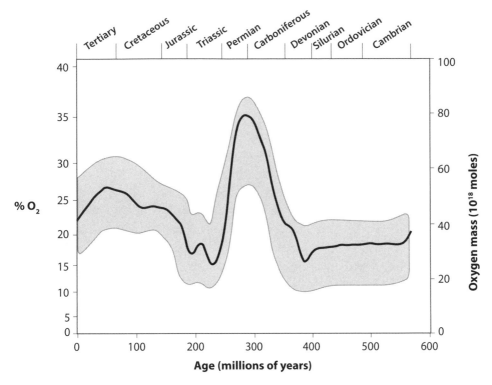

Figure 11.3. Concentrations of oxygen through the Phanerozoic Eon as calculated from rock abundance data. Redrawn from Berner and Canfield (1989), using the time scale known at the time the paper was published.

sion seen during the Carboniferous and Permian Periods. This oxygen increase was driven by the massive deposition of organic carbon in coal deposits. This, of course happened for a reason, or more likely, for a couple of reasons. The first is that land plants, which probably originated in the early Ordovician Period, diversified through the Silurian Period (see the Preface for a time scale), and by the early Carboniferous Period had grown in size and greatly expanded onto the continents. In order to stand tall, plants developed a series of tough organic molecules like lignin and cellulose, and these tend to resist microbial decay, especially when accumulated in oxygen-free environments like sediments. Thus, plant evolution played a role. The specifics of paleogeography were likely also important. During the Carboniferous and Permian Periods vast expanses of low-lying swampland collected and buried massive amounts of organic plant debris, and this is why so much coal was formed during this time. Comparable modern day environments would include the Everglades in Florida and peat bogs found in many places around the world, especially in high latitudes.

The drop in oxygen concentrations that followed was probably due to a change in paleogeography and the types of sediments deposited. Beginning in the Permian Period, but continuing into the Triassic, a supercontinent called Pangea became fully assembled. It seems that as a result of this assembly, far fewer low-lying swampy areas were available. Also, because of its vastness, less of the rain that fell onto the Pangean continent ran out to the sea, and more of it ran into the continent itself, forming vast expanses of well-drained sandy red bed deposits (we met red beds in chapter 8). These sandy continental sediments are virtually free of organic matter and thus their formation provides no input of oxygen to the atmosphere. Therefore, in Bob's model, the shift to major red bed deposition during the later Permian and Triassic Periods reduced the supply of oxygen to the atmosphere and this resulted in a precipitous drop in oxygen concentration, perhaps to levels significantly lower than in the present atmosphere.

Thus, changes in the types of sediments deposited, as controlled by a combination of factors like plant evolution, paleogeography, and climate, seem to have controlled the input rate of oxygen to the atmosphere, providing a major influence on atmospheric oxygen levels. There may even be some evidence to back at least parts of this story up. It is well known that gigantic insects were part of the Carboniferous and early Permian biosphere; some of them were almost nightmarish in size. Imagine, millipedes over a meter in length, giant spiders with leg spans as broad as an office chair (50 cm or so), and dragonflies with wingspans up to 70 cm. Indeed, these giant dragonflies prompted the French paleontologists Harlé and Harlé over a century ago to propose that atmospheric oxygen levels in the Carboniferous Period must have been higher than today's. Part of their reasoning was that higher atmospheric pressures, enabled by higher oxygen levels, would help to keep these massive fliers aloft. Also, and perhaps more importantly, they reasoned that greater oxygen availability would allow for the higher respiration rates required for such large organisms to fly, especially since they obtain their oxygen by diffusion through their tracheal systems.

These ideas, while evidently reasonable, have spurred a massive amount of debate, as well as a growing body of experimental studies. Nick Butterfield (we met Nick in chapter 10), forever the contrarian, has taken a shot at the dragonflies. He feels that ecological considerations have been a much stronger drive toward gigantism, with any assistance from higher oxygen levels taking a secondary role. One cannot overstate the importance of ecology in driving evolution, but Nick does not discuss contemporaneous gigantism among flightless insect groups. These would presumably be subject to different ecological constraints than the dragonflies. Therefore, is it simply fortuitous that a different set of ecological interactions should also drive gigantism in the flightless insect groups, or was another driver, like oxygen, in play?

There have also been a series of direct experiments on various insect groups, and these have led to equivocal results. Much of this work has been done by Jon Harrison and his group at Arizona State University. When Jon's group grew a number of different insect types under lower levels of oxygen, they usually observed a reduction in body size, a reduction in growth rates, and lower rates of survival (we will take up this topic again below). However, when grown at oxygen levels higher than those of today, mixed results were obtained. In some cases, including ground-dwelling beetles, fruit flies, and also dragonflies, body size increased at high oxygen levels. For the giant mealworm, *Zophobus morio*, body size increased at moderately elevated oxygen levels, but began to decrease again as oxygen increased further. In many other species there was no effect.

To summarize, while the results were mixed, some of experiments clearly showed increased body size at elevated oxygen levels. This result was likely a direct physiological response to higher oxygen concentrations, although larger organisms could also have been selected for in experiments running multiple generations. Nevertheless, the time scales of the experiment are probably too short to allow for the slow evolutionary changes which likely resulted in true gigantism in the fossil record. Therefore, the jury is still out, but a causal relationship between elevated oxygen levels and insect gigantism in the Carboniferous and early Permian periods remains a plausible explanation.

Back to the models. Since our publication of the history of Phanerozoic oxygen levels from rock abundance data, Bob has returned to

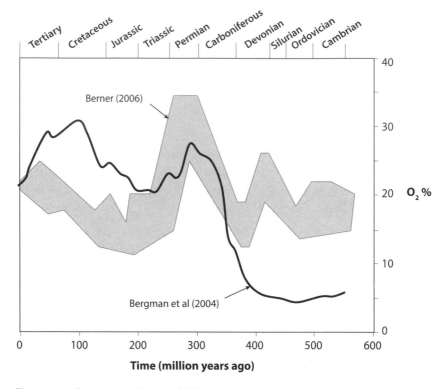

Figure 11.4. Results from Berner's GEOCARBSULF model compared to the COPSE model of Bergman and others. The grey area around the results from Berner's model encompasses the range of uncertainty as estimated by Berner, as well as newer results presented as refinements of the original GEOCARBSULF model.

oxygen modeling based on the isotope records of carbon and sulfur. This model is called GEOCARBSULF, and it is a grand model of the histories of both atmospheric CO_2 and O_2. The oxygen model results are presented in figure 11.4, and these are very similar to our earlier model results (compare to fig. 11.3), reinforcing the idea that the rock abundance and isotope data give a similar picture of the processes controlling oxygen liberation to the atmosphere.

But this isn't the end of the story. Tim Lenton from the University of Exeter, his thesis advisor Andy Watson from the University of East Anglia, and their PhD student Noam Bergman also attempted to model atmospheric oxygen (and CO_2) through Phanerozoic time, but with a somewhat different approach. Their so-called COPSE (Carbon-Oxygen-Phosphorus-Sulfur-Evolution) model is rooted in the philosophy of the

Gaia hypothesis of James Lovelock.[9] I won't get into the details here, but in its most basic form, the Gaia hypothesis posits that organisms have had an active role in shaping the chemistry of the environment. This was implicit in Bob's and my earlier model, as organisms and particularly the evolution of land plants played a major role in influencing organic carbon abundance in rocks, which influenced oxygen levels. Organisms were more explicit in Bob's recent isotope models because they directly influenced the carbon cycle in a variety of ways.[10]

Instead of relying on carbon and sulfur isotopes to fundamentally drive rates of oxygen production, the COPSE model is driven by a number of external factors, both geological and biological in nature. The factors included rates of metamorphic and volcanic outgassing, rates of tectonic uplift, land plant evolution, the enhancement of land plants on weathering, the location of organic carbon burial in the oceans (deep or shallow seas), and the increase in solar luminosity through the Phanerozoic Eon.[11]Indeed, many of these drivers can be found in Bob's models, but the COPSE model differs significantly from Bob's in some of its feedbacks. For example, the COPSE model imposes an oxygen sensitivity to the oxidation rate of sulfides and organic carbon weathering on land. This is something that the early Kump and Garrels model incorporated (recall the discussion between Karl Turekian and Bob Garrels in chapter 5), but Bob removed that sensitivity from his models long ago. The COPSE model also has a wildfire feedback (as also explored in chapter 5) to keep atmospheric oxygen concentrations from getting too high. The model also keeps track of nutrients in the oceans, mainly phosphorus, to regulate organic matter burial in sediments. In the end, rather than using the carbon and sulfur isotope curves as drivers, the COPSE model is tuned to try and recover these curves as closely as possible.[12] These curves are an integral part of the geologic record, and if they cannot be recovered, at least in their broad features, then there is likely something suspect with the model.

Results from the COPSE model are shown together with the GEO-CARBSULF model in figure 11.4. Both models have some strong similarities, particularly the increase in oxygen concentration accompanying the rise of land plants and the subsequent burial of coal in swamps, as discussed above, but there are important differences as well. The main difference is the low levels of oxygen the COPSE model produces early in the Phanerozoic Eon. In the COPSE model, this feature arises because the evolution and spread of land plants later generates a fundamental shift in oxygen regulation, giving rise to higher levels of oxygen. Thus, in the COPSE model, relatively low levels of atmospheric oxygen are the stable state before the evolution of land plants. If this is true, then the low levels of oxygen as predicted in the COPSE model in the early Phanerozoic Eon could provide an of estimate of the maximum average levels of oxygen late in the Neoproterozoic Era, as discussed in the last chapter. Thus, distinguishing between the predictions of the COPSE model and the GEOCARBSULF has important implications for unraveling the histories of atmospheric oxygen in both the late Proterozoic and the early Phanerozoic Eons.

Tais Dahl and Emma Hammarlund, whom we met in the last chapter, may have figured out a way to differentiate between these two model results.[13] If you recall, they analyzed the isotopic composition of molybdenum in sedimentary rocks deposited into ancient sulfide-rich (euxinic) environments. This approach was mentioned in the previous chapter (see also endnote 14, chapter 9), but in short, the greater the value of δ^{98}Mo, the more Mo has been removed from the oceans under oxygenated conditions, and the less has been removed into sulfidic environments like the Black Sea. In other words, to a first approximation, the greater the value of δ^{98}Mo, the more oxygenated the oceans. What Tais and Emma found was an increase in the value of δ^{98}Mo at around 400 million years ago (fig. 10.3). Sound familiar? Yes indeed, this is about the time of land plant expansion onto the continents and about the time the COPSE model suggests a major rise in atmospheric oxygen levels. This would be apparent support for the COPSE model results of the early Phanerozoic Eon.

There is more. The rise in δ^{98}Mo values is also timed with a rather profound change in the size of fish in the oceans. Indeed, we observe the emergence of fish like *Dunkleosteus*, a true monster up to 10 meters long with a thick armored skull and jaws made for crushing. As we discussed above for dragonflies, it seems reasonable that larger and more energetic fish would require more oxygen, and indeed, this seems to be true for modern fish where, on average, smaller fish can tolerate lower levels of oxygen than larger fish (fig. 11.5). One sees a similar pattern in the history of some other creatures of the sea. For example, eurypterids,

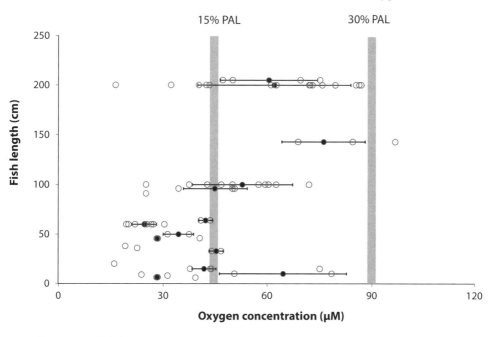

Figure 11.5. Relationship between fish size and their survival under various levels of oxygen. Plotted is the LC50, and this represents the level where 50% of the population died during a specified period of time. Averages and standard deviations are given when multiple examples of the same species were studied. Bars are also shown giving the oxygen concentrations corresponding to 15% and 30% of present atmospheric levels. Note that only small fish survive at low oxygen levels under 15% PAL. From a data compilation kindly made available by Emma Hammarlund.

better known as sea scorpions, are ancient arthropods that first appear during the Ordovician Period. Some members of this clade grew to sizes exceeding 3 meters near the Silurian-Devonian boundary, some 420 million years ago. Thus, one might argue, as Tais and Emma did, that an increase in atmospheric oxygen levels, accompanying the rise of land plants, led the way to fish (and eurypterid) gigantism in the oceans. There were, however, some big swimming organisms in the oceans before the rise of land plants. For example, there is the Middle Cambrian, meter-long *Anomalocaris*, as captured in the Burgess Shale of western Canada and in the Chengjiang deposits of China. Also, the euryptid genus *Megalogratidae* reached sizes of 1 meter in the late Ordovician Period. There were also examples of several meter-long squid-like nautiloids (in the order *Endocerida*) of the Ordovician Period some 450 million years ago. The presence of these early giants doesn't mean that Tais and Emma were wrong. Instead, it is possible that while these other animals were large, they were more sedentary with lower oxygen demand than a comparably sized fish. A more detailed look at the physiology of large and small, fast and slow, swimmers in the sea will be important in untangling these issues.

Indeed, we have plenty more to learn. The Phanerozoic Eon, however, with its rich fossil record and relative abundance of sedimentary rocks, offers the possibility for a good understanding of the history of atmospheric oxygen levels. This history can be probed by multiple lines of evidence, and it seems to encompass a beautiful relationship between the evolution of atmospheric oxygen concentrations and the evolution of life on Earth.

CHAPTER 12
Epilogue

This has been a long journey. In the end, I hope we can agree that Earth is a pretty special place. It sits at a distance from the Sun that places us in the habitable zone, allowing the persistence of liquid water. This persistence is aided by active temperature regulation, which is promoted through the relationship between atmospheric CO_2 levels, CO_2 degassing from the mantle, and the temperature control of weathering. This temperature control is also driven by plate tectonics; the same plate tectonics also drives the recycling of materials crucial to life and crucial to the liberation of oxygen into the atmosphere. Take away tectonics, and liquid water might still be found on Earth, at least in places, but life would be severely reduced in abundance. There wouldn't be the steady supply of nutrients to fuel it. The sediment cycle wouldn't take place, and organic matter and pyrite sulfur, the ultimate sources of oxygen to the atmosphere, would not be buried. Therefore, without tectonics, even if oxygen-producing organisms did exist, it's unlikely that oxygen could accumulate to appreciable amounts in the atmosphere. As it is, Earth as a planet has the "right stuff" to be an oxygen-accumulating planet. Any other planets orbiting distant stars with oxygen in their atmospheres would likely also need to have the same "right stuff." This is why NASA's search for habitable worlds is focused on finding planets within the habitable zones around their stars.

Having the right stuff, however, isn't enough. For oxygen to accumulate, it has to be produced, which means that oxygen-producing organisms must first evolve. On Earth, the pathway to the evolution of oxygen-producing cyanobacteria was apparently quite convoluted and complex. Before cyanobacteria, there were at least two different types of photosynthetic organisms that didn't produce oxygen; these were the so-called anoxygenic phototrophs. Each of these had evolved a different photosystem for converting light into energy for the cell, and the coupled photosystems in cyanobacteria emerged as an apparent fusion between the two anoxygenic phototrophic types. In making chlorophyll, cyanobacteria also borrowed, with some modifications, the pigment-producing systems of their anoxygenic phototrophic ancestors. Furthermore, the all-important oxygen-evolving complex, using a four-membered manganese cluster, may also have been borrowed, at least in broad detail, from preexisting enzymes used to convert hydrogen peroxide into water and oxygen. All in all, cyanobacteria didn't just pop up; there was a great deal of evolutionary development needed before they could arise. The famous paleontologist Steven J. Gould from Harvard was fond of discussing contingencies in evolution; in other words, the extent to which evolutionary outcomes depend on chance, like a chance encounter, a chance mutation, or chance survival under environmental stress.

The question then becomes, if we "were to play the tape again" (in Gould's words), would the outcome be the same? Would cyanobacteria have evolved in the same manner? Would coupled photosystems have emerged in the same way, or at all, or would something different have taken its place? What about the manganese cluster? These questions are perhaps more philosophical than scientific, but they are not completely trivial. In science fiction, we conjure up worlds with breathable air, and oxygen is one of the search parameters scientists use to scan for habitable (or inhabited) planets elsewhere in the galaxy and beyond. So, we imagine that the path to oxygen could and would have been followed in places other than Earth. I tend to believe that given light, water, nutrients, and time, oxygenic photosynthesis would likely evolve on other worlds. Perhaps the pathway to oxygen production would be different, but one of the remarkable aspects of life on Earth is that microbes have evolved to conduct nearly all imaginable types of metabolisms that can provide energy for growth. In this view, oxygenic photosynthesis is just one of these many possibilities.

The record on Earth suggests, however, that it takes more than the evolution of oxygen production before oxygen can accumulate in the atmosphere. There is little consensus as to when cyanobacteria evolved, but there is compelling evidence that it occurred before the "great oxidation" (the GOE) of Earth's atmosphere between 2.3 and 2.4 billion years ago. Indeed, evidence suggests that cyanobacteria existed in a largely anoxic atmosphere for hundreds of millions of years, if not a billion years or more, before the general oxygenation of the atmosphere. Our current understanding is that the same churnings of Earth that make it an immensely habitable planet, also deliver reducing gases (most importantly H_2) to the Earth surface. These gases react with oxygen, and on early Earth, they were delivered at a sufficient rate to titrate all of the oxygen liberated from the burial of organic carbon and pyrite sulfur into sediments. As Earth cooled, the tectonic churning of the planet slowed, and this resulted in a reduction in the flux of reducing gases from the mantle. With this view, the GOE marks the time in Earth history when the flux of reducing gases (again, mainly H_2) from the mantle slowed to a rate less than the rate of oxygen liberation to the atmosphere. Only then could oxygen accumulate. There is also evidence that the GOE was approached in fits and starts with occasional "whiffs" of atmospheric oxygen beginning some 300 to 400 million years earlier. To put numbers on this, the baseline oxygen levels before the GOE were probably one-thousandth of 1% of today's levels (usually called PAL, for present atmospheric levels) or less, and during the whiffs, oxygen may have risen to 0.01% to 0.1% of PAL (see figure 12.1 for a reconstruction of the oxygen history of Earth).

The GOE itself seems to have ushered in profound changes in the cycling of nutrients and carbon with surprisingly nonlinear results. First, it appears that the mobilization of nutrients, possibly phosphorus, in a newly oxygenated atmosphere accelerated organic matter production in the oceans, producing high rates of organic carbon burial and the largest positive carbon isotope excursion (the Lomagundi isotope excursion) in Earth history. It also produced a likely elevation in atmospheric oxygen levels beyond those produced during the initial stages of the GOE, perhaps even approaching modern values. A huge oxygen

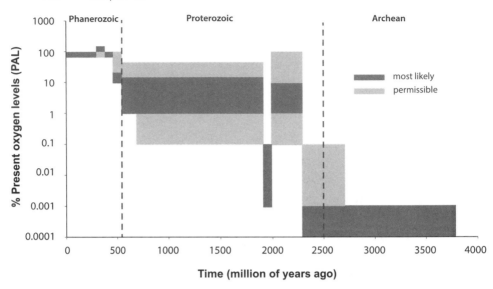

Figure 12.1. Summary of the history of atmospheric oxygen through time. See also Kump (2008)

sink was generated as this organic carbon was returned, somewhat later, into the weathering environment; this decreased oxygen to very low values, although apparently somewhat higher than those typically present before the GOE. Some 500 million years after the GOE, the carbon cycle settled into a relatively stable pattern generating oxygen levels that did not exceed 40% of PAL, and were more likely in the range of 10% to 15% of PAL or less. A lower limit on oxygen during this time has not been well established, but it is probably in the range of 0.1% to 1% of PAL. This should be enough oxygen to allow for the active weathering of organic matter and pyrite sulfur on land, which seems to have been the case.

The Neoproterozoic Era marks the transition from the dominantly microscopic life of the earlier Earth to the macroscopic life of the Phanerozoic Eon. The Neoproterozoic Era was marked by several large, perhaps global, glaciations, whose impact on the oxygen history of Earth is still uncertain. Nevertheless, a number of geochemical proxies indicate oxygenation of the oceans during the latter parts of the Neoproterozoic Era. This oxygenation is roughly correlated with the emergence of animals, including macroscopic animals with motility. Such organisms seem to require oxygen levels in the range of 10% of PAL (though many modern motile animals, especially small ones, can probably tolerate levels down to 1% of PAL), and it is tempting to conclude that the emergence of animals was enabled by increased ocean oxygenation. If this is true, then during the late Neoproterozoic Era, oxygen levels rose from values of 1% of PAL or so, to something greater, perhaps up to 10% to 15% of PAL.

In another hypothesis, oxygen levels were already in the range of 10% of PAL before the emergence of animals. With this view, large animals evolved into a permissive environment that was already in place. If this is true, then the delay in the appearance of animals is a consequence of the fact that it simply takes time for evolution to produce complex multicellular animals from simple single-celled eukaryotes. If this is the case, then the oxygenation in the oceans as revealed by geochemical measurements was not caused by an increase in the oxygen content of the atmosphere. Rather, the proxy record may be responding to an increase in the oxygenation of the oceans possibly engineered by the animals themselves. This increase in ocean oxygenation can be viewed as an expansion of the areas in the oceans where oxygen persists. I rather like this view as there are no obvious features of the carbon cycle that point to a sustained increase in oxygen liberation to the atmosphere during the late Neoproterozoic Era. More work will sort this issue out.

I think it most likely that oxygen levels during the early stages of the Phanerozoic Eon were in the range of 10% to 20% of PAL. This is consistent with some models for oxygen evolution during the Phanerozoic Eon. Geochemical measurements suggest a substantial increase in ocean oxygenation during the Late Silurian and Early Devonian Periods (some 420 million years ago), timed with a dramatic increase in the size of fishes in the sea. An explanation for this correlation is that the geochemical measurements have captured a true increase in levels of atmospheric oxygen, and extra-large fish size was enabled by the higher oxygen contents. A true increase in atmospheric oxygen levels makes some sense, because around this same time land plants evolved and expanded. The low nutrient requirements of land plants, coupled with their resistant organic matter, would have led to increased organic matter burial and enhanced oxygen liberation to the atmosphere. In this view, the evolution of land plants led to a fundamental reorganization of the carbon cycle, producing much higher levels of atmospheric oxygen, with further cascading effects on biological evolution. Due to the peculiarities of paleogeography, very high rates of organic matter burial occurred in massive swamps during the Carboniferous and early Permian Periods. This situation led to very high rates of oxygen liberation, and during this time in Earth history, atmospheric oxygen concentrations may have significantly exceeded modern levels. Insect gigantism was a visible consequence of these high oxygen levels.

As for the modern world, we are comfortably adapted to the existing levels of atmospheric oxygen. These levels are stable, and despite the burning of fossil fuels, changes in atmospheric oxygen concentration, although measurable, will be trivial until coupled, global-scale geologic/ biological processes conspire to change them. This would likely take millions of years. So, take a deep breath and ponder. There is a tremendous history behind the levels of oxygen we enjoy in the atmosphere of Earth.

NOTES

PREFACE

1. Ockham's Razor is attributed to the fourteenth-century English Friar William of Ockham. In the original form, translated from Latin, it states: "Entities should not be multiplied unnecessarily." In more pedestrian terms, Ockham's Razor states that the best explanation for something is usually the simplest explanation.

CHAPTER 1. WHAT IS IT ABOUT PLANET EARTH?

1. The redox reactions must be thermodynamically favorable, and for the aficionado, organisms actually require that the reactions are more than just thermodynamically favorable. Indeed, the reaction must be favorable by about 15 to 20 kilojoules per mole of organic carbon oxidized during a heterotrophic metabolism. This is because, at the cell level, the most basic biological function is the production of ATP. This requires the translocation of 3 to 4 protons (H^+) through an enzyme complex known as ATPase, and the minimal energy needed by an organism, therefore, is the energy needed to translocate a proton, which is estimated at around 15 to 20 kilojoules per mole of organic carbon oxidized.
2. When conditions become harsh, and/or water becomes scarce, many organisms can form spores or cysts and hold themselves in a kind of suspended animation almost indefinitely until conditions improve and they can metabolize again.
3. The name "Goldilocks Zone" refers to the traditional folk tale "Goldilocks and the Three Bears," in which Goldilocks happens upon the house of the three bears. The house is empty and upon entering, Goldilocks finds that the porridge, chair, and the bed of the little bear are "just right," and then falls asleep in the bed of the little bear. Goldilocks is awakened by little bear after the bears return, and she quickly jumps through the window in her escape from the house.
4. Albedo defines the reflectivity of an object to visible light. An object with an albedo of zero is perfectly black and absorbs all of the incoming light, whereas a perfectly reflective body is white with an albedo of 1. Thus, an object with a low albedo would absorb more light energy and become warmer than an object with a high albedo. Earth's albedo is estimated at about 0.3 including the contribution from clouds.
5. The present Venusian atmosphere contains very little water. Most of the water originally present was lost through photolysis, a light-driven process producing oxygen gas and hydrogen gas. On Venus, the hydrogen would have escaped to space, while the oxygen would have reacted either with minerals on the surface of the planet or with gases coming from within. Indeed, the present atmosphere of Venus is CO_2 rich, where the CO_2 has accumulated over eons as it degassed from the interior of the planet.
6. The Sun burns brighter with time as the ratio of H_2/He in its core decreases due to nuclear fusion reactions. According to the "standard solar model," this leads to

gravitational contraction, the release of heat, and an increase in temperature, which is expressed as in an increase in luminosity. It is estimated that when the Sun was newly formed, some 4.5 billion years ago, it was only 70% as luminous as today.

7. I come from the midwestern part of the United States, where gravestones are typically carved from limestone. Gravestones over 150 years old are often illegible, or nearly so. This is because of the weathering process. The CO_2 in the atmosphere dissolves into rain and it reacts with and slowly corrodes the limestone as the rainwater flows over its surface.

8. This evidence comes from the distribution of minerals in ancient (2.7- to 2.2-billion-year-old) soils, the so-called paleosols. These paleosols lack siderite, which, according to the calculations of Rob Rye, Phillip Kuo, and Dick Holland, means that atmospheric CO_2 levels were much lower than expected (Rye, R., Kuo, P. H., Holland, H. D., 1995. Atmospheric carbon dioxide concentrations before 2.2 billion years ago. *Nature* 378, 603–605). This topic is discussed in detail in chapter 7.

9. High methane levels might also fit with our understanding of how oxygen concentrations have changed through time, as revealed in later chapters; oxygen and methane are thermodynamically unstable together and react in the atmosphere to form CO_2.

10. Minik Rosing and colleagues argued that the albedo of early Earth was lower than is normally assumed because the continental area was less than at present, and because fewer clouds would have been generated due to fewer biologically induced cloud condensation nuclei. Lower albedo wouldn't require as high a greenhouse gas concentration in the atmosphere to warm Earth, even with a faint early Sun. These researchers also presented further evidence to support a low CO_2 atmosphere, arguing that the abundance of magnetite in Archean-aged banded iron formations is inconsistent with high levels of atmospheric CO_2 during that time. If CO_2 was much higher than at present, siderite would be the dominant Fe (iron) phase.

11. There are presently 3.16×10^{15} kg of CO_2 in the atmosphere (at 400 ppm), which is equal to 72×10^{15} moles. Total rates of photosynthesis on Earth amount to about 8×10^{15} moles y^{-1}, so the residence time of CO_2 in the atmosphere with respect to removal during photosynthesis is 72×10^{15} moles$/8 \times 10^{15}$ mol y^{-1}, or about 9 years.

12. There are presently 3.3×10^{15} moles of phosphate in the oceans. Marine rates of photosynthesis amount to about 4×10^{15} moles y^{-1} and the average C/P (carbon to phosphorous) ratio of phytoplankton is 106/1, so the residence time of P with respect to uptake by plankton is $(3.4 \times 10^{15}$ moles$/4 \times 10^{15}$ mol $y^{-1}) \times 106$, or about 86 years.

13. My geochemical colleagues will quibble with me here, because as mentioned in the text, carbon is also permanently removed in sediments as inorganic carbonate minerals. Also, bicarbonate is the major form of inorganic carbon in the oceans and its reservoir is some 50 times that of CO_2 in the atmosphere. The end result of all of this would be to lengthen the calculated residence time by a factor of 10 or so. Taking these considerations into account, however, will not change the point made in the text.

CHAPTER 2. LIFE BEFORE OXYGEN

1. As kid I had a recurring nightmare of crawling through a cave and coming to a tight spot where I stuck fast. I couldn't turn around, go forward, or backward. Creepy.

2. Water from the Colorado River is heavily used for irrigation and is a major source of drinking water to southern California. As a result, only a trickle of water now reaches the Gulf of California.

3. This is generally referred to as the photic zone.

4. These are members of the phylum *Annelida* and are therefore distant cousins to earthworms and leeches.

5. Autotrophic organisms, including autotrophic methanogens, obtain their cell material from CO_2.

6. The reaction is $4H_2 + CO_2 \rightarrow CH_4 + 2H_2O$.

7. Heterotrophs in general obtain their cell material from organic matter. Heterotrophic methanogens, in particular, are specialized in splitting acetate into methane and CO_2 (the reaction is: $H^+ + CH_2COO^- \rightarrow CH_4 + CO_2$). Acetate is a common fermentation product. Some methanogens can also decompose methyl amine compounds, thus liberating methane, and they can also produce methane from simple alcohols like methanol.

8. The reactions are:

$$SO_4^{2-} + 2CH_2O \rightarrow H_2S + 2HCO_3^-$$
$$SO_4^{2-} + 4H_2 + 2H^+ \rightarrow H_2S + 4H_2O$$

9. H_2 is a highly reduced chemical compound and therefore very well suited to reduce CO_2 to organic matter, and sulfate to sulfide, and it can also reduce many other more oxidized chemical species. Recall from the last chapter that organisms gain energy and grow by controlling these thermodynamically favorable oxidation-reduction reactions.

10. This means, quite literally, non-oxygen-producing photosynthetic organisms.

11. The reaction is: $2H_2 + CO_2 \rightarrow H_2O + CH_2O$.

12. In referring to prokaryotic organisms, the famous geobiologist Laurens Baas Becking stated in his 1934 book *Geobiologie*: "Everything is everywhere, but the environment selects." I believe that Fritz Widdel is a strong believer in this. He never seems to travel very far from the lab to collect some of the most interesting microbial species.

13. To share a few of these, in making our calculations we assumed that nutrient availability and ocean circulation rates were the same as today.

14. The name is a joke. This is an incredibly warm place.

CHAPTER 3. EVOLUTION OF OXYGENIC PHOTOSYNTHESIS

1. This and subsequent quotes from Scheele are taken as translated excerpts from Scheele's original volume. The excerpts are from: *The Discovery of Oxygen, Part 2, Experiments by Carl Wilhelm Scheele (1777)*, Alembic Club Reprints, no. 8, Edinburgh (1901) (translated by Leonard Dobbin).

2. The alchemist and inventor Cornelis Drebbel (1572–1633) apparently discovered that breathable "aerial nitre" (oxygen) could be produced by heating potassium nitrate and presumably used this technique in 1621 to support the rowing of 12 oarsman over 10 miles underwater in what was probably the first sustained submarine journey.

3. Actually 21% of our atmosphere.

4. Priestley referred to oxygen as "dephlogistated air," and did so until his death in 1804. The logic behind this name was that if the accumulation of phlogiston from burning substances caused air to lose it's ability to support a flame, then dephlogistated air would have the opposite property and support a flame.

5. This possible chain of events is explored in the play *Oxygen* by Carl Djerassi of Stanford University and Roald Hoffmann of Cornell University.

6. This quote is from Jan Ingenhousz, 1779, *Experiments upon Vegetables, Discovering Their great Power of purifying the Common Air in the Sun-shine, and of Injuring it in the Shade and at Night. To*

Which is Joined, A new Method of examining the accurate Degree of Salubrity of the Atmosphere. Printed for P. Elmsly in the Strand and H. Payne in Pall Mall, London.

7. The soluble electron carrier $NADP^+$ is an oxidized compound that accepts electrons to form NADPH, its complementary reduced form. The redox couple of $NADP^+/NADPH$ and the related $NAD^+/NADH$ are extensively used by cells to conduct oxidation-reduction reactions during cell metabolism and biosynthesis.

8. ATP is adenosine tri-phosphate. It is a high-energy compound used by cells to conduct chemical reactions that would otherwise be thermodynamically impossible. An important mode of ATP formation is through a so-called electron-transport chain, where electron transfer through a series of carrier enzymes is coupled to the transport of protons across a membrane. Protons and electrons flow back through an enzyme known as ATP synthase, forming ATP in the process.

9. Rubisco is short for ribulose 1,5 bis-phosphate carboxylase/oxygenase.

10. A small digression about GSBs. As a rule, GSBs shun oxygen and most are experts at oxidizing sulfide, while some can also oxidize Fe^{2+} (they are found in Fe^{2+}-rich Lake Matano, which was introduced in chapter 2). Some GSBs have also developed huge antenna complexes and live under ridiculously low light conditions. For example, in the Black Sea, GSBs oxidize sulfide at about a depth of 100 meters. Here, the light is equivalent to what you would see on a moonless, cloudless night in the Mohave Desert while wearing two pairs of reasonably strong sunglasses! Try it some time.

11. The anoxygenic photrophic members of this group are very diverse in lifestyle and live in environments ranging from wastewater treatment plants to sulfidic lakes. Many of these organisms can also oxidize sulfide, and some can oxidize Fe^{2+} as well. As a group, however, they are generalists, and some members can also be found in well-oxygenated environments, including the upper sunlit layers of the ocean.

12. Gene duplications are not uncommon in the history of life. A range of duplications can occur, all the way from individual genes to whole genomes. A duplicated gene, however, may evolve through mutations independently of the original gene from which it originated. In some cases (probably most cases), duplicated genes become useless and simply vanish from the genome. In other cases, the duplicated gene may share the function of the original gene, and may even end up being better than the original version. In still other cases, a duplicated gene may evolve to perform a whole new function. In general though, if the duplicated gene remains a benefit for a bacterium, it will be retained, and if it does not, it will be lost.

13. Indeed, all Rubiscos have an oxygenase activity, including so-called Rubisco-like proteins (RLPs) associated with some anaerobic *Archaea*. These RLPs are not used to fix carbon, but are believed to be the precursors to the true carbon-fixing Rubiscos.

14. One possible advantage of Rubisco is that while it has an oxygenase activity, it can still function with the Calvin cycle in the presence of oxygen, which is not true of other carbon-fixation pathways.

15. When cyanobacteria form microbial mats, oxygen concentrations within the mats may reach levels of 1 bar or higher, as explored in chapter 4. This should have also been true long ago when atmospheric oxygen levels were much lower than today.

CHAPTER 4. CYANOBACTERIA: THE GREAT LIBERATORS

1. Primary production refers to the rate of organic matter formation from CO_2 through autotrophic processes. One often refers to either gross primary production (GPP) or net primary production (NPP). Gross primary production refers to the

instantaneous rate of CO_2 fixation into organic compounds, while net primary production refers to the rate of organic matter production after cell respiration is subtracted: NPP = GPP – respiration.

2. Of course, Earth really became green with the much, much later evolution of land plants.

3. At least, a microscope would have been needed in areas other than in terrestrial hot springs where anoxygenic photosynthetic communities may have been quite visible, as explored in chapter 2.

4. In this case, the "depósita" was little more than a small wood shack where one could buy beer, soda (and sometimes ice). The beer was cheap, the service was friendly, and there were always two to three dogs lounging in the dirt just in front of the shop.

5. Perhaps most famously, modern marine stromatolites are found in Shark Bay, Australia, in Hamlin Pool, where salinity reaches about twice that of normal seawater. This salty water presumably discourages grazing animals that would otherwise eat the flourishing cyanobacteria. In general, stromatolites form as the gooey organic substances produced by cyanobacteria trap and bind solid particles, or precipitate calcium carbonate from seawater, forming solid moundlike structures.

6. This has actually been measured by Bo Barker Jørgensen of Aarhus University in Denmark. As a master of microbial mat research, Bo developed microlight probes that allowed him to explore the intensity and spectral distribution of light in microbial mats with 0.1 mm depth resolution.

7. Niels Peter Revsbech developed the so-called light-dark shift method for determining rates of oxygen production. The premise is very clever, and really quite simple. When the oxygen profile is at steady state, meaning it does not change with time, then at any given point within the profile, rates of oxygen production are balanced by rates of oxygen loss through respiration and diffusion. Niels Peter reasoned that if you turn off the lights, you remove oxygen production, but not the processes responsible for oxygen removal. Therefore, just after the lights go out, the immediate rate of oxygen decrease measured with the oxygen microelectrode is exactly the same as the rate of oxygen production just before the lights went out. In practice, to measure rates of oxygen production (equivalent to rates of primary production; see endnote 8, below) a microbial mat is darkened, and the rate of oxygen decrease is measured for a second or two. After this, the oxygen profile begins to change shape, and the rate of oxygen decrease will no longer equal the rate of oxygen production before darkening.

8. Rates of organic matter production and oxygen production by oxygenic photosynthesis are roughly equivalent. One can see this from the photosynthesis equation: $CO_2 + H_2O \rightarrow CH_2O + O_2$, where CH_2O represents organic matter.

9. Prokaryotes are what we normally think of as bacteria. They are single-celled organisms which lack, except in rare cases, organelles and a membrane-bound nucleus. Formally, they are divided into the domains *Bacteria* and *Archaea* in the tree of life.

10. This use of light energy to utilize oxygen is called the Mehler reaction.

CHAPTER 5. WHAT CONTROLS ATMOSPHERIC OXYGEN CONCENTRATIONS?

1. Oxygen is also 21% of total atmospheric gases at the top of Mount Everest, and the reduced oxygen concentration (per volume of gas) is due to the lower atmospheric pressure.

2. You might also need to add some CO_2 to the container so it doesn't start limiting the oxygen production.

3. Plants respire, like we do, to obtain energy for metabolic functions. They do this day and night, but without photosynthetic carbon production at night, there is net organic carbon oxidation. The extent to which oxygen is drawn down during the night will vary greatly from plant to plant, and this will depend mainly on the rate of plant growth. Fast growing plants will allow more oxygen accumulation than slow growing plants.

4. The organic matter could include remains of the original photosynthetically produced organic matter, or any of the organisms that have fed on this organic matter all up the food chain. These would include just about any non-photosynthetic organism ranging from heterotrophic bacteria, to fungi, fish, and horses.

5. A shale is a fine-grained, carbonate-poor, sedimentary rock composed dominantly of clay and silt-sized particles.

6. These are the relevant equations showing how pyrite burial represents an oxygen source to the atmosphere. The bottom equation is the sum of the upper three.

$16H^+ + 16HCO_3^- \rightarrow 16CH_2O + 16O_2$	Oxygenic photosynthesis
$8SO_4^{2-} + 16CH_2O \rightarrow 16HCO_3^- + 8H_2S$	Sulfate reduction
$8H_2S + 2Fe_2O_3 + O_2 \rightarrow 8H_2O + 4FeS_2$	Pyrite formation
$16H^+ + 8SO_4^{2-} + 2Fe_2O_3 \rightarrow 8H_2O + 4FeS_2 + 15O_2$	Sum

7. Both sadly are now deceased. Bob Garrels passed away in 1988, while Karl Turekian passed away just before I received the proofs of this book for correction. Both of these men were giants in geochemistry and important mentors to me in the early stages of my development as a scientist.

8. Of course, not all sediments would be prone to this type of weathering. It is imagined that a fraction of the newly deposited sediment becomes available for weathering on a short time scale (something like 10 to 20 million years), whereas the remainder becomes part of the more slowly cycling "old" rock record.

9. Subduction is a natural consequence of the plate tectonic process whereby sea-floor originally created at mid-ocean ridges is subducted back into the mantle. These are zones of mountain building and active volcanic activity. The western coast of South America is a good example of a subduction zone.

10. Anoxic means the absence of oxygen, while oxic is a state where oxygen is present. Ocean anoxia refers to a condition where parts of the ocean water column contain no oxygen. The extent of ocean anoxia has expanded and contracted through Earth history in ways discussed throughout the text.

11. Early in my career, I was quite involved in studying both the processes controlling the preservation of organic carbon in sediments and whether preservation was enhanced during anoxic decomposition. As usual, the research is a bit complicated, but the empirical evidence provided by myself and others, and subsequent experimental studies, notably by my colleague Erik Kristensen at the University of Southern Denmark, show that organic matter is more extensively decomposed with oxygen present than by anaerobic processes operating in the absence of oxygen.

12. There is, indeed, a great deal of debate around whether the availability of N or P is more important in controlling rates of primary production in the oceans. If one looks at trends of N versus P in the global ocean, in most places there is a slight phosphorus excess; this implies that nitrogen is the limiting nutrient and thus most likely to control primary production rates. The slight nitrogen deficit arises because rates of nitrogen fixation don't quite keep pace with rates of denitrification in the ocean. This can viewed as the more "biological" view on nutrient limitation. Geochemists tend to argue that

phosphorus is the limiting nutrient because the nitrogen content of the oceans should adjust itself to phosphorus content relative to the needs of the phytoplankton using the nutrients. For example, if the phosphate inventory of the oceans doubled, one might predict that nitrogen fixation would increase the nitrogen inventory to balance (or nearly so) the needs of the primary producers relative to the size of the new and larger phosphorous inventory. There is some merit to this view, and hence grounds for healthy scientific debate.

13.   The reasons for the differences in phosphorous concentrations between sediments deposited in anoxic and oxygenated waters aren't completely clear, but there are some ideas. When the water column is oxygenated, iron oxides form at the sediment surface and these strongly bind phosphate, creating a phosphorus trap within the sediments. This could be part of the reason why sediments under an oxygenated water column contain more phosphorus. But this isn't the only reason, because there is also more organic phosphorus for a given amount of organic carbon in sediments deposited under oxygenated water, compared to those deposited under euxinic conditions. Ellery Ingall has suggested that this arises because aerobic microbes may preferentially concentrate phosphorus into their biomass, and into organic phosphorus phases that are difficult to decompose, compared to the phosphorus contained in anaerobic microbes.

14.   While the fire feedback is rooted in Watson's work, and he is often given credit for it, he stated in a comment/reply on his study: "We are certainly not suggesting that the consumption of oxygen that occurs during burning has any relevance to the problem of O_2 regulation. Rather we simply point out that an oxygen concentration very much greater than the present 21% would be incompatible with the existence of a large land-based biomass because of the high fire probability" (Watson, A., Lovelock, J. E., Margulis, L., 1980. Discussion, what controls atmospheric oxygen. *Biosystems* 12, 124–125). The possible regulation of oxygen by forest fire burning was formally expressed first by Lee Kump (Kump, L. R., 1988. Terrestrial feedback in atmospheric oxygen regulation by fire and phosphorus. *Nature* 335, 152–154). Nick Lane, in his wonderful book (*Oxygen: The Molecule that Made the World*, Oxford University Press, Oxford, 2002), raises an interesting objection. He argues that the charcoal produced from fires would be very difficult to decompose and may actually increase organic carbon burial, producing more oxygen.

CHAPTER 6. THE EARLY HISTORY OF ATMOSPHERIC OXYGEN: BIOLOGICAL EVIDENCE

1.   This attribution is found in the *Metalogicon* of John of Salisbury, written in about 1159.

2.   The Vernadsky Institute of Geochemistry and Analytical Chemistry of the Russian Academy of Sciences is in Moscow, as is the Vernadsky Geological Museum; several busts and statues of his likeness can be found in the Ukraine and in Russia.

3.   Some ideas, like Ebelman's, were simply too far ahead of their time. In this case, the idea was good, it made sense, but the field was not advanced enough to do anything with it. There was not a sufficient geological context in which to place it, and thus its implications were not clear. Therefore, it was lost from the collective scientific consciousness. It was only much later, when the geological history of Earth was better understood, that the idea of oxygen control was of broad interest to the scientific community. This time it stuck when introduced by Garrels, Perry, and Holland.

4.   One is volume 19 in the Scope series, 1983, *The Global Biogeochemical Sulphur Cyle*, edited by M. V. Ivanov and J. R. Freney.

5. While sedimentary rocks are missing from this time period, some very ancient minerals can be found. For example, a particular type of mineral known as zircon ($ZrSiO_4$) has been in found in some of the oldest known sedimentary rocks, and some of these zircons have been dated as 4.3- to 4.4-billion-years-old. The rocks originally housing these zircons have been weathered away, but the zircons themselves are very resistant to weathering. Thus, they have been liberated from their parent rock through weathering to be transported in ancient rivers and deposited in younger sediments. These zircons carry some clues as to the presence or absence of water on this very early planet, and perhaps other clues that we must learn to read.

6. This redistribution is an issue we must be aware of in all rocks we look at.

7. Turbidites are a common type of sediment, generally formed in deeper water. They are formed from what you can view as an underwater avalanche when sediments from shallower depths are remobilized and flow down slope into deeper waters. Remobilization can occur because of earthquakes, for example, or from the instability inherent in sediments depositing on a steep slope.

8. Many years ago, before Minik studied his rocks, Manfred Schidlowski from the Max Plank Institute for Chemistry in Mainz, Germany, looked at and measured the isotopic composition of graphite in Isua rocks. These, however, were not the same rocks looked at by Minik, and they did not have the same geological context. His values were generally less ^{13}C-depleted than those found by Minik. (For an explanation of these carbon isotopes, see the main text.)

9. $\delta^{13}C = 1000(R13/12_{sample} - R13/12_{standard})/(R13/12_{standard})$, where R13/12 is the ratio of carbon-13 to carbon-12 in either our sample or our standard.

10. This symbol ‰ means per mil, which is the same as parts per thousand. In comparison, percent is parts per hundred. Therefore, the per mil value is 10 times larger than the percent value.

11. There are ways of producing organic matter by abiological means with isotope signals similar to those found in Isua. However, these reactions require catalysts that are not apparent in the Isua rocks. Furthermore, such inorganic pathways, which are unknown in the marine water column, cannot easily account for the interbedding of organic layers with turbidites in the Isua rocks.

12. Some cyanobacteria make certain cells called heterocysts, where the nitrogen fixation we encountered in chapter 4 takes place; some other cyanobacteria produce resting cells called akinetes. Both akinetes and heterocycts have distinctive morphologies and if found in a filament, they provide compelling evidence for cyanobacteria. In other cases, the morphology of ancient, well-preserved cyanobacteria resembles modern species so closely that a cyanobacterial interpretation is difficult to escape.

13. Some of these would be filamentous sulfur bacteria of the genera *Beggiatoa* and *Thioploca*, which oxidize sulfide with oxygen and nitrate to survive.

14. Very recently, Martin Brasier and colleagues presented some evidence for fossil bacteria in rocks some 60-million-years younger than the Apex Chert by (Wacey, D., Kilburn, M. R., Saunders, M., Cliff, J., Brasier, M. D., 2011. Microfossils of sulphur-metabolizing cells in 3.4-billion-year-old rocks of Western Australia. *Nature Geoscience* 4, 698–702). These fossil forms have carbon isotope signatures consistent with life. The fossils themselves are not terribly well preserved, and they will likely be subject to the same type of scrutiny as the Apex Chert fossils described by Bill Schopf, but the carbon isotope evidence seems a pretty sound indicator of life.

15. They also found a variety of so-called hopane molecules, and in particular, 2-methyl hopanes. It used to be thought that these were specific cyanobacterial biomarkers, but subsequent work has shown that they are also produced by noncyanobac-

terial prokaryotes (Welander, P. V., Coleman, M. L., Sessions, A. L., Summons, R. E., Newman, D. K., 2010. Identification of a methylase required for 2-methylhopanoid production and implications for the interpretation of sedimentary hopanes. *PNAS* 107, 8537–8542). Consequently, they are no longer viewed as strict cyanobacterial biomarkers.

16. Eukaryotes are organisms whose cells contain a membrane-bound nucleus and organelles like mitochondria, and include chloroplasts in plants.

CHAPTER 7. THE EARLY HISTORY OF ATMOSPHERIC OXYGEN: GEOLOGICAL EVIDENCE

1. Sadly, Dick passed away after I wrote the first version of this chapter. I had, however, sent him a copy as soon as the draft was finished, and I can only hope that he enjoyed the opening pages.

2. The Paleozoic is the first of the three eras in the Phanerozoic Eon, which is the age of animals. The Phanerozoic Eon extends from the end of the Proterozoic Eon, 542 million years ago, until the end of the Permian Period, 251 million years ago. The end of the Permian Period is marked by the largest known extinction of animal life in Earth history. See the Preface and figure P.1 for more details.

3. In fact, much of the gold was redistributed by later hydrothermal activity, and although there is still active debate, one school of thought holds that the redistributed gold was derived from original detrital gold grains. There seems to be some evidence for this as some detrital gold grains are still apparent.

4. We could make excursions to various places in the United States, Canada, South Africa, India, the Ukraine, Brazil, and Greenland and observe similar rocks spanning the Archean Eon in age. There are also some younger examples of these types of rocks, from ages concentrated around 1.9 billion years ago and 600 to 700 million years ago, and these will be discussed in later chapters.

5. James Farquhar is very modest, and I'm sure he didn't quite put it this way.

6. The isotope ^{32}S has 16 protons and 16 neutrons, while ^{33}S has the same number of protons and 17 neutrons. The isotope ^{34}S has 18 neutrons, and there are 20 neutrons for ^{36}S.

7. Because of the low sulfate concentration, gypsum will not form as seawater concentrates, so we don't need to worry about gypsum as a riverine sulfate source in a low oxygen world.

8. We are talking here about the redox state of the mantle. This can be assessed, for example, by looking at the concentrations of vanadium in volcanic rocks. Vanadium partitions itself into a partial melt of mantle rocks depending on the redox state of the mantle where the melt formed. Therefore, when these melts erupt onto Earth's surface, the vanadium contents provide a measure of the redox state of the mantle. If we look at such volcanic rocks through time, we can assess the evolution of mantle redox state, and available evidence suggests that it hasn't changed much (see for example, Canil, D., 2002. Vanadium in peridotites, mantle redox and tectonic environments: Archean to present. *Earth and Planetary Science Letters* 195, 75–90).

9. The idea here is that to generate a mass-independent isotope effect from a single starting compound, say SO_2, that initially displays a mass-dependent isotope signature, one must generate at least two products showing opposite mass-independent isotope signals. Imagine now that subsequent atmospheric chemistry, in this case oxidation with

oxygen, causes each of the products to form a single compound, sulfate in this case. The sulfate will have formed by combining the opposite mass-independent isotope signals from the original compounds formed from the photodissociation of SO_2, generating a product with the same mass-dependent isotope signal as the original SO_2.

CHAPTER 8. THE GREAT OXIDATION

1. Actually, Cloud preferred the age range of 2.0 to 2.2 billion years ago, based on the available chronologies at the time. Subsequent and better dating has pushed the ages back somewhat.

2. Dick Holland also looked at the chemistry of ancient soils, the so-called paleosols. These are remnants of old soil horizons preserved in the geologic record. To make a long story short, before the GOE (great oxidation event), soils lost a great deal of iron through weathering. This is because in a low-oxygen atmosphere, the weathering process solubilized iron from the rock, but as there was no oxygen to oxidize it as rust from solution, dissolved iron (Fe^{2+}) was transported from the soil into streams and rivers and presumably into the oceans. After the GOE, iron reacted with oxygen and was then retained as oxidized minerals in the soil (see, for example, Rye, R., Holland, H. D., 1998. Paleosols and the evolution of atmospheric oxygen: A critical review. *American Journal of Science* 298, 621–672).

3. One of Joe Kirschvink's ideas is the "Snowball Earth" hypothesis. Here, Joe recognized that many glacial deposits associated with widespread glaciations during the Neoproterozoic Era were located near the Equator. Somewhat earlier, Mikhail Budyko from Russia had conducted heat-budget modeling of the Earth surface and concluded that if glacial ice ever extended to within about 50° of the Equator, the albedo would be so high that Earth could not retain sufficient heat from the Sun to keep the glaciers from covering the globe. This is known as a runaway icehouse, and Joe applied this model to the Neoproterozoic glaciations, arguing that during this era, Earth entered into a runaway icehouse, possibly several times, freezing solid and generating a "Snowball Earth." He also had clever suggestions as to how Earth escaped this condition.

4. Carbon dioxide (CO_2), carbonic acid (H_2CO_3), bicarbonate ion (HCO_3^-), and carbonate ion (CO_3^{2-}) are all related to one another through a chemical equilibrium governed by pH. At the pH of seawater, about 8.0, the bicarbonate ion dominates; at a lower pH, carbonic acid and CO_2 become more important; and at a higher pH, the carbonate ion gains prominence.

5. Although this isotope spike was first described by Manfred Schidlowski from the Max Planck Institute for Chemistry in Mainz, Germany, in the 1970s.

6. This experiment is borrowed from *How the Weather Works*, by Michael Allaby (Dorling Kindersley, London, 1995). Completely fill a clear glass with colored hot water (but not boiling hot) and cover the glass with aluminum foil. Put the covered glass into a large clear vase or aquarium with uncolored cold water and either carefully remove or poke out the aluminum foil cover. See what happens. Compare this to what happens when the cup and tank contain water of the same temperature, or have only slightly different temperatures.

7. Planetesimals are small solid objects formed during the early developmental stages of the solar system. Through violent collisions, these planetesimals formed into larger bodies and eventually, in our region of the solar system, into Earth.

8. Plotted as H_2 demand, which is twice the rate of O_2 production: $2H_2 + O_2 \rightarrow 2H_2O$.

9. Dick's calculations are fundamentally based on the isotope record of organic carbon and inorganic carbon and what these records imply about the burial rates of organic carbon through time. To make these estimates one also needs to know how the size of the carbon reservoir at the Earth surface has changed through time. These ideas are taken from the work of John Hayes and Jake Waldbauer who published a wonderful paper on the evolution of the carbon cycle through time (Hayes, J. M., Waldbauer, J. R., 2006. The carbon cycle and associated redox processes through time. *Philosophical Transactions of the Royal Society B* 361, 931–950). The idea is that when Earth was young, there was less carbon at the surface, and the carbon inventory could only grow as CO_2 degassed from the mantle. Hayes and Waldbauer estimated this growth rate as well as the burial fluxes of organic carbon through time, and Holland used these results in his model.

CHAPTER 9. EARTH'S MIDDLE AGES:
WHAT CAME AFTER THE GOE

1. What is now the surface of the shale was once found deep within the rock. As the subsequent layers have been eroded and weathered away, the surface as we see it now has become exposed. But, on its way to becoming the top, this part of the shale also experienced weathering and oxygen exposure while it was within the rock. So, what represents the surface of the rock now, has also experienced all the intermediate stages of weathering that the buried parts of the rock are now undergoing.

2. This was modeled by John Hayes and Jake Waldbauer as discussed in endnote 9 of the previous chapter.

3. This is an oversimplification of course. There were the pre-GOE "whiffs" of oxygen discussed in Chapter 7, and some oxidation of pyrite and other oxygen-sensitive species occurred during these times. Just how significant this oxidation was, however, is uncertain.

4. The reaction is: $14H_2O + 4FeS_2 + 15O_2 \rightarrow 4Fe(OH)_3 + 8SO_4^{2-} + 16H^+$.

5. And it is a somewhat tautological argument as it turns out. The late Precambrian evolution of animals is sometimes taken as evidence for a late Precambrian rise in oxygen, and sometimes the evolution of animals is attributed to a late Precambrian rise in oxygen. You can't have it both ways, without some independent evidence for a late Precambrian rise in oxygen levels.

6. This has to do with the specifics of how sulfate reducers transport sulfate into the cells and the biochemical pathways leading to fractionation within the cells. Thus, sulfur isotope fractionation occurs through at least two steps within the cells, both involving the breaking of S-O bonds. In the first step, the sulfate ion (SO_4^{2-}) is reduced to the sulfite ion (SO_3^{2-}) (by the enzyme APS reductase), and in the second step, sulfite is reduced to sulfide (by the enzyme sulfite reductase). The fractionations imparted by these steps, however, can only be expressed if the cell actively exchanges with the sulfate outside the cell. If not, all the sulfate entering the cell is reduced to sulfide and no fractionation will be observed despite the enzyme-driven fractionations within the cell. As sulfate concentration becomes lower, sulfate becomes limiting to the cell and is exchanged less with the external environment. In this way, lower sulfate concentrations lead to lower fractionations.

7. Rob was an integral part of my graduate student experience. He visited every summer while I was working on my PhD, and we became good friends and colleagues.

We shared lots of laughs, lots of beers, and lots of arguments about the geochemistry of modern and ancient marine sediments. Our collaboration has continued to today.

8.   Upwelling refers to the physical transport of water from deeper to shallower regions in the water column of the ocean. It is often driven by wind, which can, for example, push surface water away from a coast. Deeper water is then drawn up to the replace the displaced surface water. This happens today in many regions of the ocean, but conspicuously along the western coast of North and South America.

9.   Dick Holland once calculated that oxygen levels a thousand times less than present-day levels would be needed for Fe^{2+} to be transported for deposition in 2.5 billion year old BIFs from South Africa; these also formed in shallow water.

10.   This evidence was provided by my colleague Robert Frei of the University of Copenhagen. The short story is that Cr (chromium) isotopes are fractionated during the oxidative weathering of Cr minerals on land. This oxidative weathering requires oxygen, but Mn oxides are the direct oxidant, whose formation requires oxygen. Once transported into the oceans, the isotopic signal of Cr is captured in iron-rich rocks like BIFs. The record shows some episodes of fractionated Cr after about 2.6 billion years ago, yielding more evidence for oxygen "whiffs" before the GEO. However, there is no evidence for fractionated Cr in the 1.88 billion year old Gunflint Iron Formation, suggesting very low oxygen at this time, which will be discussed later in the chapter.

11.   Indeed, Lee Kump from Penn State University has recently presented evidence which may support this idea. He showed that the isotopic composition of both inorganic carbon and organic matter in the aftermath of the Lomagundi excursion became way too ^{13}C-depleted to be explained by the normal operation of the carbon cycle, and he appealed to the massive input of ^{13}C-depleted carbon from the oxidation of organic matter accumulated on the continents (Kump, L. R., Junium, C., Arthus, M. A., Brasier, A., Fallick, A., Melezhik, V., Lepland, A., Crne, A. E., Luo, G. M., 2011. Isotopic evidence for massive oxidation of organic matter following the Great Oxidation Event. *Science* 334, 1694–1696).

12.   But not exclusively so. Some of the blackest shales I've seen come from the Neoproterozoic Era, and most of these contain very little pyrite and were likely deposited under Fe-rich, ferruginous water column conditions.

13.   In practice, we focused on understanding how Fe was partitioned in the samples, and then used this to determine the chemical environment in which the samples were deposited. If the samples contained excess Fe over what is expected in normal sediments depositing from a water-column rich in oxygen, this means that the water column overlying the sediments was oxygen-free, and therefore capable of transporting dissolved Fe^{2+} in the water and ultimately to the sediment. If most of the Fe enrichment was bound as pyrite, this means that the overlying water contained sulfide. If the enriched Fe was pyrite poor, then sulfide was low in concentration and the overlying water contained Fe^{2+} instead. The overlying water could not have been both high in sulfide and in Fe^{2+}, due to the low solubility of iron sulfide minerals.

14.   Yikes, another isotope system! As explored in chapter 7, molybdenum (Mo) is liberated during oxidative weathering on land and is then transported by rivers to the ocean. There are two major removal pathways in the ocean. One is by adsorption onto Fe and Mn oxides, representing the oxic removal pathway, and here a rather big fractionation is incurred. There is little fractionation during removal into sulfidic environments, which is the second removal pathway. Therefore, sediments deposited in sulfidic settings provide an estimate of the isotopic composition of Mo in ancient seawater, and this isotopic composition provides a measure of the balance between oxic and sulfidic removal pathways. So, Gail Arnold and colleagues presented Mo isotope evidence for an enhanced sulfide removal pathway of Mo in ancient 1.6 billion

year old sediments from Northern Australia, which were themselves deposited in a sulfidic setting.

15. The Fe^{2+} would have come from a combination of hydrothermal inputs from deep-sea vents and from rivers where much of the iron would have been bound as iron oxides coating river particulates. These immobile Fe oxides can be reduced to the soluble ferrous state under anoxic conditions.

CHAPTER 10. NEOPROTEROZOIC OXYGEN
AND THE RISE OF ANIMALS

1. Letter to King Charles I, 19 August, 1628. Manuscript, CO 1/5 (27), 75, MHA 16-B-2-011, transcribed by P. E. Pope, with spelling and punctuation modernized. From Public Record Office, Colonial Office, London, England, 1999. Documents Relating to Ferryland: 1597–1726. Colony of Avalon History, Colony of Avalon Foundation, Ferryland, NL, and Newfoundland and Labrador Heritage Web Site Project, Memorial University of Newfoundland, St. John's, NL, Canada.

2. Epithelial cells line surfaces and cavities in animals and act as a protective covering, facilitate exchange and allow organisms to detect the environment.

3. The Ediacaran Fauna, while highly visible and easily recognizable in the fossil record, may not represent the earliest evidence of fossil animal life. Gordon Love from the University of California, Riverside, has identified sponge-associated biomarkers in rocks of about 635 million years of age (Love, G. D., et al., 2009. Fossil steroids record the appearance of Demospongiae during the Cryogenian period. *Nature* 457, 718–721); and Adam Maloof from Princeton University has found some possible sponge body fossils from slightly older rocks (Maloof, A. C., et al., Possible animal-body fossils in pre-Marinoan limestones from South Australia. *Nature Geoscience* 3, 653–659).

4. The Gaskiers glaciation was possibly the last, and perhaps also the smallest of the major glaciations to punctuate the latter half of the Neoproterozoic Era. Some of these glaciations were so large that they nearly enclosed Earth in ice, generating the so-called Snowball Earth. The Snowball Earth hypothesis was a dream child of Joe Kirschvink's; we met Joe in chapter 8. The snowball Earth hypothesis is more fully discussed in endnote 3, chapter 8, and was made famous by Paul Hoffman and his colleagues in a remarkable paper from 1998 (Hoffman, P. F., Kaufman, A. J., Halverson, G. P., Schrag, D. P., 1998. A Neoproterozoic Snowball Earth. *Science*. 281, 1342–1346). These glaciations and their potential influence on life and global chemistry are actively discussed in the literature, but very little consensus has yet been reached. I have elected not to focus on these glaciations here, but a nice introduction can be found in Andy Knoll's book *Life on a Young Planet* (Princeton University Press, Princeton, NJ, 2003).

5. We collected with proper permissions and not from the famous fossil beds!

6. In a geologic context, "deep ocean" normally means anything below storm wave base, or below about 100 meters depth. Not that deep really. In the case of the Avalon rocks, independent evidence suggests that we are probably looking at depths much greater: hundreds of meters or perhaps even a kilometer. Unless we have a piece of deep seafloor emplaced (obducted) onto the continents, we never get a look at the real deep ocean.

7. The others contributing to this work included Andy Knoll, Guy Narbonne, Gerry Ross, Tatiana Goldberg, and Harald Strauss.

8. This first evidence is not to be confused with the first occurrence. There could have been earlier episodes of deep ocean oxygenation that we have missed.

9.   Fully air-saturated water in the modern world contains an amount of oxygen that depends on the temperature and salinity of the water. The bottom water of the ocean is derived from polar regions where the air-saturated concentration is 325 micromoles per liter. The water in lakes, rivers, and coming out of your faucet at home is probably not too far from this value.

10.   In an earlier contribution, Andreas Teske, now at the University of North Carolina, and I came up with a similar estimate of minimum oxygen levels. This was based on the amount of oxygen required to oxygenate surface marine sediments, because we showed that an oxidative sulfur cycle was active in sediments and this required oxygen.

11.   If you rip a piece of paper in half, the area around the margins of the paper has increased. Rip each piece in half again, and the area has increased even further.

12.   Lou looked at the ratio of $^{87}Sr/^{86}Sr$ in the strontium preserved in limestones. If unaltered by later processes, the isotopic composition of Sr in seawater, which is incorporated into limestones as they precipitate, indicates the balance between input of strontium from the continents, which has a high ratio of $^{87}Sr/^{86}Sr$, and input from hydrothermal vents, which has a low ratio of $^{87}Sr/^{86}Sr$. The limestone ratio increases dramatically at around 600 million years ago, indicating a greater input of continental weathering products into the ocean, and therefore, a high burial rate of sediment with associated organic matter. Lou, however, didn't construct an oxygen model, and one of my concerns is that a high consumption rate of oxygen associated with high rates of continental weathering would have balanced the high rates of oxygen liberation associated with the high rates of organic carbon burial. This point has been made by Bob Berner, and it will be explored in more detail in chapter 11.

13.   Not all with agree with this, including Martin Kennedy from the University of Adelaide, and also Lou Derry. They argue that the anomaly is a diagenetic effect caused by the percolation of fluids through the limestones long after original deposition, thus changing the original isotopic compositions. One cannot rule out that diagenetic effects have influenced the limestones, particularly their ^{18}O compositions. However, the ^{13}C signal is observed at many sites worldwide and may extend over hundreds of kilometers at individual locations. I cannot think of a plausible diagenetic mechanism which can produce such broad scale and reproducible ^{13}C features, so I favor the interpretation that these represent original seawater compositions.

14.   We argued that the release and oxidation of a giant pool of methane tied up in sediments as clathrate hydrate phases may be a better solution. These are found today in marine sediments and are even more ^{13}C depleted than the dissolved organic carbon (DOC) found in the oceans.

15.   Somewhat less of an oxygen sink is required to produce the Shuram-Wonoka anomaly by methane oxidation when compared to DOC oxidation. This is because methane is much more ^{13}C-depleted than marine DOC. Indeed, Christian Bjerrum and I have modeled the anomaly with methane oxidation while still preserving enough oxygen for early animal respiration.

16.   Clearly, the earliest evolution of primitive animals occurred much earlier than this and was likely not influenced by the higher oxygen demands of motile organisms. Indeed, preliminary work in our group suggests that 2% of present oxygen levels may be sufficient for the respiration of sponges, which is the most basal of the modern animals in the tree of life.

CHAPTER 11. PHANEROZOIC OXYGEN

1. There actually are windows, but they are small slits included more for design effect than to give anyone a sense of light and of life outside of the building.

2. We have also seen this ability in Vladamir Vernadsky, Preston Cloud, and Dick Holland, all mentioned earlier in the text.

3. I won't get into this in detail, but rates of weathering should depend directly on the area of continent available for weathering. These rates should also depend on the activity of the hydrologic cycle, and on temperature, with the higher the temperature the faster the rate of weathering. In this case, more rain leads to more rapid weathering. The greater the relief of the continents, that is the height of the mountains relative to the valleys, the more intense the weathering, particularly physical weathering and erosion. Also, plants, and their root systems physically disrupt rocks and soils and they also elevate CO_2 levels in the soil; these influences combine to increase the rates of weathering.

4. Ronov was a great inspiration to the western geological community, even during the Soviet Era when contact between western and Soviet scientists was difficult at best.

5. They are low in sulfur because sulfate concentrations are low in terrestrial waters like in lakes and rivers. Low sulfate concentrations limit the amount of sulfate reduction and therefore the amount of sulfide that can be produced to react with either iron to form pyrite, or with organic matter to form organic sulfur compounds.

6. The rate of deposition of an individual sediment type is calculated as the proportion of the total sediment represented by an individual rock type multiplied by the total sediment deposition rate; for example, if coal deposited during a particular time represents 10% of the total sediment, the multiplier is 0.1.

7. This is because both the weathering of organic carbon and pyrite sulfur, and their subsequent burial again in marine sediments, are coupled to the total rate of sediment deposition. The processes are coupled because weathering on land generates the sediment that deposits in the sea. Furthermore, the burial rates of organic carbon and pyrite sulfur are a linear function of sedimentation rate. Thus, since both the liberation and consumption of oxygen are governed by total sediment deposition rate, the variability of this rate has little influence on the net accumulation or consumption of oxygen from the atmosphere. See note 12 in the previous chapter.

8. A similarity is also found between calculated rates of sulfur burial from the sediment abundance data and those calculated from sulfur isotopes.

9. Lovelock introduced the Gaia hypothesis in the 1970s (Lovelock, J. E., Margulis, L., 1974. Atmospheric homeostasis by and for the biosphere: The Gaia hypothesis. *Tellus* Series A 26, 2–10; and Lovelock, J. E., 1979. *Gaia: A New Look at Life on Earth*, Oxford University Press, Oxford), and in its original form, Lovelock envisioned a global biosphere acting in harmony to generate a chemical environment that was optimized for life. Thus, life controlled its environment. As it became clear that Earth surface chemistry has changed through time, the idea of an invariant optimization became relaxed, but the idea that life has contributed heavily to shaping, or even controlling, the chemical environment has persisted. Indeed, the Gaia hypothesis has much in common with the earlier writings of Vladimer Vernadsky, as discussed in chapter 6. Few working geobiologists would refute the importance of life in influencing the chemical environment. The role of life in actively regulating this chemistry is a frontier research area and a topic of intense discussion.

10. One of these ways is through photorespiration. If you recall from chapter 3, the Rubisco enzyme is responsible for carbon fixation in the Calvin cycle and also has an oxygenase activity (consuming O_2 with eventual release of CO_2), which competes

directly with its carboxylase function (fixing CO_2 to form organic carbon). This means that the enzyme becomes less efficient at fixing carbon at higher oxygen levels as the oxygenase activity becomes more important. You could view this as an organism-level biological negative feedback on oxygen increase, although most evidence suggests that the oxygenase activity is an evolutionary quirk rather than a decided biological innovation for oxygen stabilization. In any event, Bob recognized the importance of this as an oxygen regulator and as a feedback mechanism in his oxygen model.

11.   If you remember back to chapter 1, solar luminosity was a big deal in considering the temperature regulation of early Earth; it is less significant, but still of some importance, in the Phanerozoic Eon.

12.   These isotope curves can be recovered because the COPSE model does compute the ancient cycles of carbon and sulfur, including the burial rates of organic carbon and pyrite into sediments. Thus, the isotopic composition of all the elements of the ancient carbon and sulfur cycles can be predicted.

13.   There may also be another way to look at this. As introduced in chapter 5, the incidence of wildfires is oxygen sensitive. Experimental burning of plant material, coupled with modeling, suggests that plants will not burn at oxygen levels below 16%, and that burning becomes rampant above 22% (the present atmosphere is 21% O_2) (Belcher, C. M., Yearsley, J. M., Hadden, R. M, McElwain, J. C., Rein, G., 2010. Baseline intrinsic flammability of Earth's ecosystems estimated from paleoatmospheric oxygen over the past 350 million years. *PNAS* 107, 22448–22453). Charcoal is formed when plants burn, and the geologic record of charcoal remains suggests that fires occurred as early as the latest Silurian, about 418 million years ago (Scott, A. C., Glasspool, I. J., 2006. The diversification of Paleozoic fire systems and fluctuations in atmospheric oxygen concentration. *PNAS* 103, 10861–10865). Elevated oxygen levels of 16% at this time would be compatible with the molybdenum isotope evidence, and consistent with GEOCARBSULF models. The COPSE model predicts a somewhat later rise in oxygen to these levels, but it could be made consistent with the charcoal results with different assumptions about the timing of the development of the terrestrial land-plant ecosystem.

REFERENCES

PREFACE

Gradstein, F., Ogg, J., Smith, A., 2004. *A Geologic Time Scale*. Cambridge University Press, Cambridge, UK.

Hutton, J., 1788. *Theory of the Earth*. Royal Society of Edinburgh, Edinburgh. (Classic book and many would argue it marked the beginning of modern geology.)

CHAPTER 1

Baross, J. A., Benner, S .A., Cody, G. D., Copley, S .D., Pace, N. R., Scott, J. H., Shapiro, R., Sogin, M. L., Stein, J. L., Summons, R. E., Szostak, J. W., 2007. The Limits of Organic Life in Planetary Systems. National Academy of Sciences, Washington, DC. (National Academy of Sciences reports on the limits of life. See chapter 6 for the importance of water, and speculations on other possible solvents for life.)

Canfield, D. E., Kristensen, E., Thamdrup, B., 2005. *Aquatic Geomicrobiology*. Academic Press, Amsterdam. (See discussion of redox reactions and life in chapter 3.)

Harland, D. M., 2005. *Water and the Search for Life on Mars*. Springer, Berlin. (A popular book outlining evidence for water on Mars up to and including early results from the Mars Exploration Rover [MER] Mission.)

Kasting, J. F., 1993. Habitable zones around main sequence stars. *ICARUS* 101, 108–128. (Excellent discussion of the habitability zone, including a nice historical perspective.)

Kasting, J. F., 2010. *How to Find a Habitable Planet*. Princeton University Press, Princeton, NJ. (Entertaining and readable account of the evolution of Earth's atmosphere and climate, and the search for habitable worlds elsewhere in the solar system.)

Knoll, A. H., 2003. *Life on a Young Planet. The First Three Billion Years of Evolution on Earth*. Princeton University Press, Princeton, NJ, and Oxford, UK. (A fantastic journey through the first 3 billion years of life on Earth.)

Mustard, J. F., Murchie, S. L., Pelkey, S. M., Ehlmann, B. L., Milliken, R. E., et al., 2008. Hydrated silicate minerals on mars observed by the Mars reconnaissance orbiter CRISM instrument. *Nature* 454, 305–309. (Spectroscopic evidence for hydrated silicate minerals, implying weathering in an aqueous fluid.)

Rosing, M. T., Bird, D. K., Sleep, N. H., Bjerrum, C. J., 2010. No climate paradox under the faint early Sun. *Nature* 464, 744–747. (Rosing and colleagues argue that because of lower albedo on early Earth, there was in fact no faint young Sun paradox.)

Sagan, C., Mullen, G., 1972. Earth and Mars: Evolution of atmospheres and surface temperatures. *Science* 177, 52–56. (Sagan and Mullen introduce the faint young Sun paradox.)

Squyres, S. W., Arvidson, R. E., Bell, J. F., Calef, F., Clark, B. C., et al., 2012. Ancient impact and aqueous processes at Endeavour Crater, Mars. *Science* 336, 570–576. (Recent evidence for water from the MER mission.)

Walker, J.C.G., Hays, P. B., Kasting, J. F., 1981. A negative feedback mechanism for the long-term stabilization of Earth's surface temperature. *Journal of Geophysical Research-Oceans and Atmospheres* 86, 9776–9782. (A negative feedback is introduced through the carbon cycle to solve the faint young Sun paradox.)

CHAPTER 2

Baas Becking, L.G.M., 1925. Studies on the sulphur bacteria. *Annals of Botany* 39, 613–650. (The idea of the sulfuretum is first presented.)

Baas Becking, L.G.M., 1934. *Geobiologie: of Inleiding tot de Milieukunde*. W.P. Van Stockum & Zoon N. V., Den Haag, The Netherlands. (Classic book defining "geobiology" as a field.)

Brock, T. D., 1994. *Life at High Temperatures*. Yellowstone Association for Natural Science, History & Education, Yellowstone National Park, WY. (Popular introduction into life in hydrothermal springs with a focus on Yellowtone National Park. Written by Thomas Brock, the pioneer in the field.)

Canfield, D. E., Rosing, M. T., Bjerrum, C., 2006. Early anaerobic metabolisms. *Philosophical Transactions of the Royal Society B* 361, 1819–1834. (An attempt to constrain the activity level of the biosphere before oxygenic photosynthesis.)

Crowe, S. A., Jones, C., Katsev, S., Magen, C., O'Neill, A. H., Sturm, A., Canfield, D. E., et al., 2008. Photoferrotrophs thrive in an Archean Ocean analogue. *Proceedings of the National Academy of Sciences of the United States of America* 105, 15938–15943. (A potential ancient ocean analogue with Fe-rich waters likely supporting Fe-oxidizing anoxygenic phototrophs.)

David, L. A., Alm, E. J., 2011. Rapid evolutionary innovation during an Archaean genetic expansion. *Nature* 469, 93–96. (Awesome look at the evolution of metabolic pathways.)

Kharecha, P., Kasting, J., Siefert, J., 2005. A coupled atmosphere-ecosystem model of the early Archean Earth. *Geobiology* 3, 53–76. (A clever use of atmospheric-ocean models to explore the activity of an early-Earth biosphere.)

Shen, Y., Buick, R., Canfield, D. E., 2001. Isotopic evidence for microbial sulphate reduction in the early Archean era. *Nature* 410, 77–81. (Earliest evidence for microbial sulfate reduction.)

Shen, Y. N., Farquhar, J., Masterson, A., Kaufman, A. J., Buick, R., 2009. Evaluating the role of microbial sulfate reduction in the early Archean using quadruple isotope systematics. *Earth and Planetary Science Letters* 279, 383–391. (Further evidence for the early evolution of microbial sulfate reduction.)

Ueno, Y., Yamada, K., Yoshida, N., Maruyama, S., Isozaki, Y., 2006. Evidence from fluid inclusions for microbial methanogenesis in the early Archaean era. *Nature* 440, 516–519. (Evidence for the early evolution of methanogenesis.)

Widdel, F., Schnell, S., Heising, S., Ehrenreich, A., Assmus, B., Schink, B., 1993. Ferrous iron oxidation by anoxygenic phototrophic bacteria. *Nature* 362, 834–835. (First description of anoxygenic phototrophic oxidation of Fe^{2+}.)

CHAPTER 3

Allen, J. F., 2005. A redox switch for the origin of two light reactions in photosynthesis. *FEBS Letters* 579, 963–968. (Here, Allen lays out his hypothesis for the origin of the coupled reaction centers in oxygenic phototrophs.)

Allen, J. F., Martin, W., 2007. Out of thin air. *Nature* 445, 610–612. (Nice overview of the evolution of oxygenic photosynthesis with an interesting teaser as to the development of the manganese cluster.)

Badger, M. R., Bek, E. J., 2008. Multiple Rubisco forms in proteobacteria: Their functional significance in relation to CO_2 acquisition by the CBB cycle. *Journal of Experimental Botany* 59, 1525–1541. (Excellent review of Rubisco, its multiple forms, and its catalytic ability.)

Blankenship, R. E., 1992. Origin and early evolution of photosynthesis. *Photosynthesis Research* 33, 91–111. (Original papers describing the evolution of the PSI and PSII from the reaction centers used among anoxygenic phototrophs.)

Blankenship, R. E., 2010. Early evolution of photosynthesis. *Plant Physiology* 154, 434–438. (Excellent review of photosynthesis evolution.)

Blankenship, R. E., Hartman, H., 1998. The origin and evolution of oxygenic photosynthesis. *Trends in Biochemical Sciences* 23, 94–97. (Describes the evolution of oxygenic photosynthesis and the oxygen evolving complex from catylase enzymes.)

Canfield, D. E., Kristensen, E., Thamdrup, B., 2005. *Aquatic Geomicrobiology.* Academic Press, Amsterdam.. (See chapter 4 for a review of photosynthesis.)

Falkowski, P. G., Raven, J. A., 2007. *Aquatic Photosynthesis*. Princeton Unversity Press, Princeton, NJ. (THE book on all aspects of aquatic photosynthesis.)

Hohmann-Marriott, M. F., Blankenship, R. E., 2011. Evolution of photosynthesis, *Annual Review of Plant Biology* 62, 515–548. (Recent review on the evolution of photosynthesis.)

Raymond, J., Blankenship, R. E., 2008. The origin of the oxygen-evolving complex. *Coordination Chemical Reviews* 252, 377–383. (Structural study on the similarity between the oxygen evolving complex and Mn catylase.)

Sadekar, S., Raymond, J., Blankenship, R. E., 2006. Conservation of distantly related membrane proteins: Photosynthetic reaction centers share a common structural core. *Molecular Biology and Evolution* 23, 2001–2007. (Study comparing all of the structures of the photosynthetic reaction centers.)

Tabita, F. R., Hanson, T. E., Li, H. Y., Satagopan, S., Singh, J., Chan, S., 2007. Function, structure, and evolution of the RubisCO-like proteins and their RubisCO homologs. *Microbiology and Molecular Biology Reviews* 71, 576–599. (Excellent review of Rubisco and Rubisco-like proteins.)

Xiong, J., Fischer, W. M., Inoue, K., Nakahara, M., Bauer, C. E., 2000. Molecular evidence for the early evolution of photosynthesis. *Science* 289, 1724–1730. (Molecular study on the evolution of photosynthetic pigment formation pathways.)

CHAPTER 4

Berman-Frank, I., Lundgren, P., Chen, Y.-B., Küpper, H., Kolber, Z., Bergman, B., Falkowski, P., 2001. Segregation of nitrogen fixation and oxygenic photosynthesis in the marine cyanobacterium *Trichodesmium*. *Nature* 294, 1534–1537.

(Brilliant study on the regulation of nitrogen fixation by the most important marine nitrogen fixer.)

Budel, B., Weber, B., Kuhl, M., Pfanz, H., Sultemeyer, D., Wessels, D., 2004. Reshaping of sandstone surfaces by cryptoendolithic cyanobacteria: Bioalkalization causes chemical weathering in arid landscapes. *Geobiology* 2, 261–268. (Nice review of endolithic cyanobacteria including where they live and their influence on the rock.)

Canfield, D. E., Des Marais, D.J., 1991. Aerobic sulfate reduction in microbial mats. *Science* 251, 1471–1473. (Description of sulfate reduction under oxic conditions in a microbial mat and a description of microbial mat structure.)

Canfield, D. E., Des Marais, D. J., 1993. Biogeochemical cycles of carbon, sulfur, and free oxygen in a microbial mat. *Geochimica et Cosmochimica Acta* 57, 3971–3984. (Description of carbon and sulfur cycles in a microbial mat ecosystem.)

Canfield, D. E., Kristensen, E., Thamdrup, B., 2005. *Aquatic Geomicrobiology*. Academic Press. Amsterdam. (See chapter 7 for a review of the nitrogen cycle and chapter 13 for a review of cyanobacterial microbial mats.)

Chisholm, S. W., Olson, R. J., Zettler, E. R., Goericke, R., Waterbury, J. B., Welschmeyer, N. A., 1988. A novel free-living prochlorophyte abundant in the oceanic euphotic zone. *Nature* 334, 340–343. (First description of *Prochlorococcus* in the oceans.)

Curtis, S. E., Clegg, M. T., 1984. Molecular evolution of chloroplast DNS-sequences. *Molecular Biology and Evolution* 1, 291–301. (Early molecular evidence linking the chloroplast to cyanobacteria.)

Des Marais, D. J., 1990. Microbial mats and the early evolution of life. *Tree* 5, 140–144. (Nice early review of microbial mats and their relevance for studies of early Earth enviroments.)

Garcia-Pichel, F., Belnap, J., Neuer, S., Schanz, F., 2003. Estimates of global cyanobacterial biomass and its distribution. *Algological Studies* 109, 213–227. (Quantitative review of the significance of cyanobacteria in a variety of ecosystems and on a global scale.)

Johnson, P. W., Sieburth, J. M., 1979. Chroococcoid cyanobacteria in the sea: Ubiquitous and diverse phototropic biomass. *Limnology and Oceanography* 24, 928–935. (First description of the marine cyanobacteria that turned out to be *Prochlorococcus*.)

Johnson, Z. I., Zinser, E. R., Coe, A., McNulty, N. P., Woodward, E.M.S., Chisholm, S. W., 2006. Niche partitioning among Prochlorococcus ecotypes along ocean-scale environmental gradients. *Science* 311, 1737–1740. (Beautiful description of the partitioning of various *Prochlorococcus* strains with depth in the oceans, and as a function of latitude.)

Jørgensen, B. B., Des Marais, D. J., 1988. Optical properties of benthic photosynthetic communities: Fiber-optic studies of cyanobacterial mats. *Limnology and Oceanography* 33, 99–113. (Early study of the optical properties of microbial mats, made possible with the development of special fiber-optic microsensors.)

Kettler, G. C., Martiny, A. C., Huang, K., Zucker, J., Coleman, M. L., Rodrigue, S., Chen, F., et al., 2007. Patterns and implications of gene gain and loss in the evolution of Prochlorococcus. *PLOS Genetics* 3, 2515–2528. (Nice discussion of the plasticity of the *Prochlorococcus* genome.)

Khakhina, L. N., 1992. *Concepts of Symbiogenesis: Historical and Critical Study of the Research of Russian Scientists*. Yale University Press, New Haven, CT. (Summary of the contributions of Konstantin Sergeevich Merezhkovsky and other Russian scientists to the development of the endosymbiotic theory for eukaryotic cell development.)

Liu, H., Nolla, H. A., Campbell, L., 1997. *Prochlorococcus* growth rate and contribution to primary production in the equatorial and subtropical North Pacific Ocean. *Aquatic Microbial Ecology* 12, 39–47. (Estimates of the contribution of *Prochlorococcus* to marine primary production.)

Revsbech, N. P., 1989. An Oxygen Microsensor with a guard cathode. *Limnology and Oceanography* 34, 474–478. (Description of the most widely used oxygen microelectrode in ecological research.)

Revsbech, N. P., Jørgensen, B. B., Blackburn, T. H., Cohen, Y., 1983. Microelectrode studies of the photosynthesis and O_2, H_2S, and pH profiles of a microbial mat. *Limnology and Oceanography* 28, 1062–1074. (Brilliant study of oxygen and sulfur dynamics in microbial mats using newly developed microelectrodes.)

Sagan, L., 1967. On the origin of mitosing cells. *Journal of Theoretical Biology* 14, 225–274. (First statement by Lynn Margulis, then Lynn Sagan, on the endosymbiotic theory for the origins of chloroplasts and mitochondria.)

Severin, I., Stal, L. J., 2010. Spatial and temporal variability in nitrogenase activity and diazotrophic community composition in coastal microbial mats. *Marine Ecology-Progress Series* 417, 13–25. (Study showing a variety of different strategies used by benthic nitrogen-fixing communities.)

Sohm, J. A., Webb, E. A., Capone, D. G., 2011. Emerging patterns of marine nitrogen fixation. *Nature Reviews Microbiology* 9, 499–508. (Nice review of nitrogen fixation in the oceans.)

Waterbury, J. B., Watson, S. W., Guillard, R.R.L., Brand, L. E., 1979. Widespread occurrence of a unicellular, marine, planktonic, cyanobacterium. *Nature* 277, 293–294. (First description of *Synecococcus* in the oceans.)

Zobell, C. E., 1946. *Marine Microbiology*. Chronica Botanica Company, Waltham, MA. (Classic book on the microbiology of the seas.)

CHAPTER 5

Berner, R. A., 1987. Models for carbon and sulfur cycles and atmospheric oxygen: Application to Paleozoic geologic history. *American Journal of Science* 287, 177–196. (Bob's first oxygen regulation paper where he introduced the concept of rapid recycling.)

Berner, R. A., 2004. *The Phanerozoic Carbon Cycle: CO2 and O2*. Oxford University Press, Oxford, UK. (Comprehensive treatment of the long-term modeling of CO_2 and O_2 in the atmosphere, based to a large extent on Bob Berner's own work.)

Berner, R. A., Maasch, K. A., 1996. Chemical weathering and controls on atmospheric O_2 and CO_2: Fundamental principles were enunciated by J.J. Ebelman in 1845. *Geochimica et Cosmochimica Acta* 60, 1633–1637. (Historical account of the role of Ebelman in elucidating the controls on atmospheric oxygen concentrations.)

Canfield, D. E., 2005. The early history of atmospheric oxygen: Homage to Robert M. Garrels. *Annual Review of Earth and Planetary Science* 33, 1–36. (A survey of the history of atmospheric oxygen, with an entrance point in the work of Bob Garrels.)

Ebelman, J. J., 1845. Sur les produits de la décomposition des especes minérales de la famille des silicates. *Annales des Mines* 7, 3–66. (First account of the controls on atmospheric oxygen concentrations.)

Fennel, K., Follows, M., Falkowski, P. G., 2005. The co-evolution of the nitrogen, carbon and oxygen cycles in the Proterozoic ocean. *American Journal of Science* 305, 526–545. (A model study of the role of nitrogen in controlling marine primary production during periods of Earth history when the deep oceans were largely anoxic.)

Garrels, R. M., Perry, E. A., 1974. Cycling of carbon, sulfur, and oxygen through geologic time, in: Goldberg, E. D. (ed.), *The Sea*. John Wiley and Sons, New York, pp. 303–336. (Classic modern account of the cycling of elements and processes controlling oxygen concentrations.)

Holland, H. D., 1978. *The Chemistry of the Atmosphere and Oceans*. John Wiley and Sons, New York. (Classic text on the processing that acts to control the chemistry of the atmosphere and the oceans.)

Ingall, E., Jahnke, R., 1994. Evidence for enhanced phosphorus regeneration from marine sediments overlain by oxygen depleted waters. *Geochimica et Cosmochimica Acta* 58, 2571–2575. (Evidence is presented supporting Ingall's original idea that phosphorus is preferentially regenerated from sediments underlying an oxygenated water column.)

Kump, L. R., 1988. Terrestial feedback in atmospheric oxygen regulation by fire and phosphorus. *Nature* 335, 152–154. (The role of fire in regulating atmospheric oxygen levels is taken seriously.)

Murray, J. W., Codispoti, L. A., Friederich, G. E., 1995. Oxidation-reduction environments: The suboxic zone in the Black Sea. *Advances in Chemistry Series* 244, 157–176. (Nice review of Black Sea water-column chemistry.)

Petsch, S. T., Berner, R. A., Eglinton, T. I., 2000. A field study of the chemical weathering of ancient sedimentary organic matter. *Organic Geochemistry* 31, 475–487. (Beautiful field study of the oxidative weathering of organic carbon and sulfur during soil formation.)

Richards, F. A., 1965. Anoxic basins and fjords, in: Riley, J. P., Skirrow, G. (eds.), *Chemical Oceanography*. Academic Press, London, pp. 611–645. (Classic account of the chemistry of anoxic basins and fjords.)

Van Cappellen, P., Ingall, E. D., 1996. Redox stabilization of the atmosphere and oceans by phosphorus-limited marine productivity. *Science* 271, 493–496. (Modeling study showing how the phosphorus feedback of Ingall can work to regulate oxygen levels.)

Watson, A., Lovelock, J. E., Margulis, L., 1978. Methanogenesis, fires and the regulation of atmsopheric oxygen. *Biosystems* 10, 293–298. (First study of the influence of oxygen concentrations on burning.)

CHAPTER 6

Brasier, M. D., Green, O. R., Jephcoat, A. P., Kleppe, A. K., van Krankendonk, M. J., Lindsay, J. F., Steele, A., Grassineau, N. V., 2002. Questioning the evidence for Earth's oldest fossils. *Nature* 416, 76–81. (This paper marks Martin Brasier's first assult on the biogenicity of Bill Schopf's 3.45 billion-year-old microfossils.)

Brasier, M. D., Green, O. R., Lindsay, J. F., McLoughlin, N., Steele, A., Stoakes, C., 2005. Critical testing of earth's oldest putative fossil assemblage from the similar to 3.5 Ga Apex Chert, Chinaman Creek, Western Australia. *Precambrian Research* 140, 55–102. (This paper is a response to Schopf's reply to Brasier's original critique of Schopf's fossils. Here Brasier uses, among other things, raman spectroscopy to argue that the "fossils" of the Apex Chert are nonbiological in origin.)

Rosing, M. T., 1999. ^{13}C-depleted carbon microparticles in >3700-Ma sea-floor sedimentary rocks from West Greenland. *Science* 283, 674–676. (Minik Rosing presents carbon isotope evidence for life in the ancient sedimentary rocks of Isua, Greenland.)

Schopf, J. W., 1993. Microfossils of the early Archean Apex Chert: New evidence of the antiquity of life. *Science* 260, 640–646. (Bill Schopf present his iconic images of what he interpreted as possible cyanobacterial microfossils from the 3.45 billion year old Apex Chert in Western Australia.)

Schopf, J. W., Kudryavtsev, A. B., 2009. Confocal laser scanning microscopy and Raman imagery of ancient microscopic fossils. *Precambrian Research* 173, 39–49. (Schopf presents more laser raman evidence that structures from the Apex Chert are ancient microbial fossils.)

Schopf, J. W., Kudryavtsev, A. B., 2012. Biogenicity of Earth's earliest fossils: A resolution of the controversy. *Gondwana Research* 22, 761–771. (Schopf and Kudryavtsev provide further evidence for the biogenicity of the Apex Chert fossils.)

Schopf, J. W., Kudryavtsev, A. B., Agresti, D. G., Wdowiak, T. J., Czaja, A. D., 2002. Laser-Raman imagery of Earth's earliest fossils. *Nature* 416, 73–76. (Schopf uses raman spectroscopy to rebut Brasier's arguments that the structures Schopf originally descibed in the Apex Chert are nonbiogenic.)

Summons, R. E., Bradley, A. S., Jahnke, L. L., Waldbauer, J. R., 2006. Steroids, triterpenoids and molecular oxygen. *Philosophical Transactions of the Royal Society B* 361, 951–968. (A great review of the role of oxygen in sterol synthesis.)

Van Kranendonk, M. J., 2006. Volcanic degassing, hydrothermal circulation and the flourishing of early life on Earth: A review of the evidence from c. 3490–3240 Ma rocks of the Pilbara Supergroup, Pilbara Craton, Western Australia. *Earth-Science Reviews* 74, 197–240. (Includes a review of the evidence for the hydrothermal origin of the rocks containing the Apex Chert fossils.)

Vernadsky, V. I., 1998. *The Biosphere*. Springer-Verlag New York, New York. (Classic book arguing for the role of life in shaping the geosphere and the atmosphere.)

Waldbauer, J. R., Newman, D. K., Summons, R. E., 2011. Microaerobic steroid biosynthesis and the molecular fossil record of Archean life. *Proceedings of the National Academy of Sciences of the United States of America* 108, 13409–13414. (Impressive study showing that only trace amounts of oxygen are needed for steroid sysnthesis by aerobes.)

Waldbauer, J. R., Sherman, L. S., Sumner, D. Y., Summons, R. E., 2009. Late Archean molecular fossils from the Transvaal Supergroup record the antiquity of microbial diversity and aerobiosis. *Precambrian Research* 169, 28–47. (Beautiful study of biomarkers including steranes in 2.46 to 2.67 billion year old rocks, and their relationship to oxygen.)

CHAPTER 7

Anbar, A. D., Duan, Y., Lyons, T. W., Arnold, G. L., Kendall, B., Creaser, R. A., Kaufman, A. J., et al., 2007. A whiff of oxygen before the Great Oxidation Event? *Science* 317, 1903–1906. (A beautiful look at the geochemistry of the 2.5-billion-year-old Mt. McRae shale, showing evidence for mild oxidation of Earth's atmosphere before the GOE. The term oxygen "whiff" is coined here.)

Farquhar, J., Bao, H. M., Thiemens, M., 2000. Atmospheric influence of Earth's earliest sulfur cycle. *Science* 289, 756–758. (Game-changing paper showing the common occurrence of a mass-independent sulfur isotope signal before the GOE, and the disappearance of it afterwards. The best available evidence for low atmospheric oxygen before the GOE.)

Farquhar, J., Savarino, J., Airieau, S., Thiemens, M. H., 2001. Observation of the wavelength-sensitive mass-dependent sulfur isotope effects during SO_2 photolysis: Implications for the early atmosphere. *Journal of Geophysical Research* 106, 32829–32839. (Clever photochemical experiments linking the mass-independent sulfur isotope signal to photochemical reactions with UV radiation.)

Frimmel, H. E., 2005. Archaean atmospheric evolution: Evidence from the Witwatersrand gold fields, South Africa. *Earth-Science Reviews* 70, 1–46. (Modern review on placer uraninite and pyrite in the Witwatersrand mines, and their relevanace to the evolution of atmospheric evolution.)

Holland, H. D., 1962. Model for the evolution of the Earth's atmosphere, in: Engel, A.E.J., James, H. L., Leonard, B. F. (eds.), *Petrologic Studies: A Volume in Honor of A. F. Buddington*. Geological Society of America, Boulder, CO, pp. 447–477. (Amazing paper laying out Dick Holland's early thinking on the evolution of atmosphereic oxygen concentrations. This paper laid the foundation for much of his future work.)

Holland, H. D., 1984. *The Chemical Evolution of the Atmosphere and Oceans*. Princeton University Press, Princeton, NJ. (Classic book outlining Dick's views on the evolution of atmospheric and ocean chemistry as of 1984. Still widely used.)

Holland, H. D., 2009. Why the atmosphere became oxygenated: A proposal. *Geochimica et Cosmochimica Acta* 73, 5241–5255. (Brilliant paper offering a plausible explanation for the GOE.)

Pavlov, A. A., Kasting, J. F., 2002. Mass-independent fractionation of sulfur isotopes in Archean sediments: Strong evidence for an anoxic Archean atmosphere. *Astrobiology* 2, 27–41. (Atmospheric chemical model calibrating the oxygen levels needed to create mass-independent sulfur isotope fractionations.)

Phillips, G. N., Law, J.D.M., Myers, R. E., 2001. Is the redox state of the Archean atmosphere constrained? *SEG Newsletter* 47, 8–18. (A skeptic's view on the detrital origin of Witwatersrand uraninites and pyrites.)

Rasmussen, B., Buick, R., 1999. Redox state of the Archean atmosphere: Evidence from detrital heavy minerals in ca. 3250–2750 Ma sandstones from the Pilbara Craton, Australia. *Geology* 27, 115–118. (Presents more occurrences of detrital pyrite and uraninite during the Archean Eon.)

Utter, T., 1980, Rounding of ore particles from the Witwatersrand gold and uranium deposit (South Africa) as an indicator of their detrital origin. *Journal of Sedimentary Petrology* 71–76. (Nice SEM and petrographic study of detrital uraninite and pyrite grains from the Witwatersrand.)

Wille, M., Kramers, J. D., Nagler, T. F., Beukes, N. J., Schroder, S., Meisel, T., Lacassie, J. P., Voegelin, A. R., 2007. Evidence for a gradual rise of oxygen between 2.6 and 2.5 Ga from Mo isotopes and Re-PGE signatures in shales. *Geochimica et Cosmochimica Acta* 71, 2417–2435. (Presents the first geochemical evidence for a "whiff" of oxygen before the GOE.)

CHAPTER 8

Catling, D. C., Claire, M. W., 2005, How the Earth's atmosphere evolved to an oxic state: A status report. *Earth and Planetary Science Letters* 237, 1–20. (A nice summary of the controls on atmospheric oxygen regulation, and explanations for the GOE, as of 2005.)

Catling, D. C., Zahnle, K. J., McKay, C. P., 2001. Biogenic methane, hydrogen escape, and the irreversible oxidation of early life. *Science* 293, 839–843. (A very clever model with which Catling and colleagues explain the GOE through the accumulated effect of the photolysis of methane to H_2, and the loss of H_2 to space, leading to the oxidation of the surface environment.)

Cloud, P. E., 1968. Atmospheric and hydrospheric evolution on the primitive earth. *Science* 160, 729–736. ("Big-thinking" paper that outlines the major stages in the chemical and biological evolution of Earth.)

Cloud, P. E., Jr., 1972. A working model of the primitive Earth. *American Journal of Science* 272, 537–548. (Builds on the ideas presented in the 1968 paper.)

Crowell, J. C., 1995. Preston Cloud, 1912–1991: A biographical memoir, in: *Biographical Memoirs*. National Academy of Sciences Press, Washington, DC, pp. 43–63. (Lovely biography of Preston Cloud.)

Guo, Q. J., Strauss, H., Kaufman, A. J., Schroder, S., Gutzmer, J., Wing, B., Baker, M. A., et al., 2009. Reconstructing Earth's surface oxidation across the Archean-Proterozoic transition. *Geology* 37, 399–402. (Refines the timing of the GOE through S isotope evidence in Huronian-aged sedimentary rocks.)

Holland, H. D., 1994. Early Proterozoic atmospheric change, in: Bengston, S. (ed.), *Early Life on Earth*. Columbia University Press, New York, NY, pp. 237–244. (A nice summary of evidence for the GOE.)

Holland, H. D., 2009. Why the atmosphere became oxygenated: A proposal. *Geochimica et Cosmochimica Acta* 73, 5241–5255. (Holland "titration" proposal for the GOE, where O_2 rises when the flux of reduced gases from the mantle falls below the rate of O_2 liberation through organic carbon and pyrite burial.)

Karhu, J. A., Holland, H. D., 1996. Carbon isotopes and the rise of atmospheric oxygen. *Geology* 24, 867–870. (A compilation of carbon isotope evidence demarking the Lomagundi isotope event, originally throught to be the cause of the GOE.)

Kasting, J. F., Eggler, D. H., Raeburn, S. P., 1993. Mantle redox evolution and the oxidation state of the archean atmosphere. *Journal of Geology* 101, 245–257. (In a clever proposal, Kasting and colleagues propose that the mantle was more reducing on early Earth, allowing a reducing atmosphere even in the face of oxygen production by cyanobacteria.)

Kirschvink, J. L., Kopp, R. E., 2008. Palaeoproterozoic ice houses and the evolution of oxygen-mediating enzymes: The case for a late origin of photosystem II. *Philosophical Transactions of the Royal Society B* 363, 2755–2765. (Here, Kirschvink and Kopp make the case that the GOE was a result of the evolution of cyanobacteria at around 2.3 billion years ago, and they discuss and dismiss evidence for cyanobateria predating this time.)

Kump, L. R., Kasting, J. F., Barley, M. E., 2001. Rise of atmospheric oxygen and the "upside-down" archean mantle. *Geochemistry Geophysics Geosystems* 2, paper number 2000GC000114. (Kump and colleagues link the GOE to mantle overturn, where volcanic gases originate from more oxidized source regions than they did previously when the atmosphere was reducing.)

Sekine, Y., Suzuki, K., Senda, R., Goto, K. T., Tajika, E., Tatad, R., Goto, K., et al., 2011, Osmium evidence for synchronicity between a rise in atmospheric oxygen and Paleoproterozoic degalciation. *Nature Communications* 2. doi:10.1038/ncomms1507. (Nice study providing evidence for osmium mobility as a result of the GOE.)

CHAPTER 9

Arnold, G. L., Anbar, A. D., Barling, J., Lyons, T. W., 2004. Molybdenum isotope evidence for widespread anoxia in mid-Proterozoic oceans. *Science* 304, 87–90. (First paper showing Mo isotopes could constrain the extent of euxinic conditions in ancient oceans.)

Bekker, A., Slack, J. F., Planavsky, N., Krapez, B., Hofmann, A., Konhauser, K. O., Rouxel, O. J., 2010. Iron formation: The sedimentary product of a complex interplay among mantle, tectonic, oceanic, and biospheric processes. *Economic Geology* 105, 467–508. (A nice contemporary review of banded iron formations.)

Cameron, E. M., 1982. Sulphate and sulphate reduction in early Precambrian oceans. *Nature* 296, 145–148. (Found high sulfur isotope fractionations after the GOE and postulated a rise in seawater sulfate concentrations.)

Canfield, D. E., 1998. A new model for Proterozoic ocean chemistry. *Nature* 396, 450–453. (Suggested that oxygenation of the atmosphere drove the oceans to become more sulfidic, causing a massive removal of dissolved iron, which called an end to BIF deposition.)

Canfield, D. E., 2005. The early history of atmospheric oxygen: Homage to Robert M. Garrels. *Annual Review of Earth and Planetary Science* 33, 1–36. (Review of the history of atmospheric oxygen and of the processes regulating its concentration.)

Canfield, D. E., Poulton, S. W., Knoll, A. H., Narbonne, G. M., Ross, G., Goldberg, T., Strauss, H., 2008. Ferruginous conditions dominated later Neoproterozoic deep-water chemistry. *Science* 321, 949–952. (Documentation of widespread ferruginous deep-water marine conditions in the Neoproterozoic Era.)

Canfield, D. E., Teske, A., 1996. Late Proterozoic rise in atmospheric oxygen concentration inferred from phylogenetic and sulphur-isotope studies. *Nature* 382, 127–132. (Sulfur isotope evidence for a Neoproterozoic rise in atmospheric oxygen concentrations.)

Frei, R., Gaucher, C., Poulton, S. W., Canfield, D. E., 2009. Fluctuations in Precambrian atmospheric oxygenation recorded by chromium isotopes. *Nature* 461, 250–253. (Use of Cr isotopes to document fluctations in the oxygenation of the oceans, and in particular, the very low oxygen conditions following the Lomagundi isotope excursion.)

Holland, H. D., 2002. Volcanic gases, black smokers, and the great oxidation event. *Geochimica et Cosmochimica Acta* 66, 3811–3826. (Here, Dick Holland provides a possible link between the GOE and the Lomagundi isotope excursion based on the availability of phosphorus.)

Holland, H. D., 2004. The geologic history of seawater, in: Holland, H. D., Turekian, K. K. (eds.), *Treatise on Geochemistry*. Elsevier, Amsterdam, pp. 583–625. (Excellent, wide-ranging, review, but here in particular, Holland calculated how the presence of Fe^{2+} in the surface waters over a broad continental shelf constrains atmospheric oxygen concentrations.)

Planavsky, N. J., McGoldrick, P., Scott, C. T., Li, C., Reinhard, C.T., Kelly, A.E., Chu, X., et al., 2011. Widespread iron-rich conditions in the mid-Proterozoic ocean. *Nature* 477, 488–451. (Demonstration of widespread ferruginous conditions through the middle Proterozoic Eon.)

Poulton, S. W., Canfield, D. E., Fralick, P., 2004. The transition to a sulfidic ocean ~1.84 billion years ago. *Nature* 431, 173–177. (Demonstration of a transition from BIF to euxinic conditions following the deposition of the Gunflint Iron Formation.)

Poulton, S. W., Fralick, P. W., Canfield, D. E., 2010. Spatial variability in oceanic redox structure 1.8 billion years ago. *Nature Geoscience* 3, 486–490. (Documents the 2-D structure of ocean chemistry during and after the deposition of the Gunflint Iron Formation. Shows that euxinic conditions extended some 100 km from the coast, where ferruginous conditions then developed.)

Raiswell, R, Canfield, D. E., 2012, The iron biogeochemical cycles past and present. *Geochemical Perspectives* 1, 1–220. (The biochemical cycling of Fe past and present, as related through the careers of Raiswell and Canfield.)

Sarmiento, J. L., Herbert, T. D., Toggweiler, J. R., 1988. Causes of anoxia in the world ocean. *Global Biogeochemical Cycles* 2, 115–128. (Simple and beautiful model of the controls on deep-water marine oxygenation.)

Shen, Y., Knoll, A. H., Walter, M. R., 2003. Evidence for low sulphate and anoxia in a mid-Proterozoic marine basin. *Nature* 423, 632–635. (Showed the presence of long-standing deep-water sulfidic conditions in an approximately 1.5 billion year old marine basin from Northern Territory, Australia.)

Slack, J. F., Grenne, T., Bekker, A., Rouxel, O. J., Lindberg, P. A., 2007. Suboxic deep seawater in the late Paleoproterozoic: Evidence from hematitic chert and iron formation related to seafloor-hydrothermal sulfide deposits, central Arizona, USA. *Earth and Planetary Science Letters* 255, 243–256. (Shows that marine deep waters were not sulfidic in the late Paleoproterozoic Eon.)

Wilkinson, B. H., McElroy, B. J., Kesler, S. E., Peters, S. E., Rothman, E. D., 2009. Global geologic maps are tectonic speedometers: Rates of rock cycling from area-age frequencies. *Geological Society of America Bulletin* 121, 760–779. (Modern analysis of the preservation of rocks as a function of time.)

CHAPTER 10

Billings, E., 1872. On some fossils from the Primordial rocks of Newfoundland. *The Canadian Naturalist* 4. 465–479. (First description of Ediacaran fauna.)

Bjerrum, C. J., Canfield, D. E., 2011. Towards a quantitative understanding of the late Neoproterozoic carbon cycle. *Proceedings of the National Academy of Sciences of the United States of America* 108, 5542–5547. (A model to explain large Neoproterozoic negative isotope excursions based on the non-steady-state workings of the methane cycle.)

Butterfield, N. J., 2011. Animals and the invention of the Phanerozoic Earth system. *Trends in Ecology and Evolution* 26, 81–87. (A view as to how animals have shaped the geochemical environment.)

Canfield, D. E., Poulton, S. W., Knoll, A. H., Narbonne, G. M., Ross, G., Goldberg, T., Strauss, H., 2008. Ferruginous conditions dominated later Neoproterozoic deep water chemistry. *Science* 321, 949–952. (Paper showing the widespread development of deep-water ferruginous conditions during the Neoproterozoic Era.)

Canfield, D. E., Poulton, S. W., Narbonne, G. M., 2007. Late-Neoproterozoic deep-ocean oxygenation and the rise of animal life. *Science* 315, 92–95. (Paper showing the oxygenation of the deep ocean of the Avalon Peninsula around 580 million years ago.)

Dahl, T. W., Hammarlund, E. U., Anbar, A. D., Bond, D.P.G., Gill, B. C., Gordon, G. W., Knoll, et al., 2010. Devonian rise in atmospheric oxygen correlated to the radiations of terrestrial plants and large predatory fish. *Proceedings of the National Academy of Sciences of the United States of America* PNAS 107, 17911–17915. (Molybdenum isotope evidence is used to support a Neoproterozoic rise in atmospheric oxygen concentrations as well as a later rise in the Devonian Period.)

Derry, L. A., Jacobsen, S. B., 1990. The chemical evolution of precambrian seawater: Evidence for REEs in banded iron formation. *Geochimica et Cosmochimica Acta* 54, 2965–2977. (Clever application of multiple isotope systems to calculate rates of organic carbon burial during the Neoproterozoic Era.)

Knoll, A. H., 2011. The multiple origins of complex multicellularity. *Annual Review of Earth and Planetary Sciences* 39, 217–239. (Fascinating exploration of evolution of multicellularity among differet groups or organisms with a wonderful discussion of the early evolution of animals.)

Knoll, A. H., Hayes, J. M., Kaufman, A. J., Swett, K., Lambert, I. B., 1986. Secular variation in carbon isotope ratios from Upper Proterozoic successions of Svalbard and East Greenland. Nature 321, 832–838. (First carbon isotope record indicating high rates of organic carbon burial during the Neoproterozoic Era.)

Narbonne, G. M., 2005. The Ediacara biota: Neoproterozoic origin of animals and their ecosystems. *Annual Review of Earth and Planetary Science* 33, 421–442. (Excellent review of the Ediacaran Fauna.)

Nursall, J. R., 1959. Oxygen as a prerequisite to the origin of the metazoa. *Nature* 183, 1170–1172. (Early discussion of the relationship between the history of atmospheric oxygen concentrations and animal evolution.)

Rothman, D. H., Hayes, J. M., Summons, R. E., 2003. Dynamics of the Neoproterozoic carbon cycle. *Proceedings of the National Academy of Sciences of the United States of America* 100, 8124–8129. (Clever analysis of large Neoproterozoic Era carbon isotope excursions with the suggestion that they originated from the periodic oxidation of a huge pool of marine dissolved organic carbon.)

Runnegar, B., 1982. Oxygen requirements, biology and phylogenetic significance of the late Precambrian worm *Dickinsonia*, and the evolution of the burrowing habit. *Alcheringa* 6, 223–239. (An attempt to place limits on atmospheric oxygen levels at the dawn of animal life, based on calculations regarding the physiological requirements of the Ediacaran fossil *Dickinsonia*.)

Seilacher, A., 1992. Vendobionta and Psammocorallia: Lost constructions of Precambrian evolution. *Journal of the Geological Society of London* 149, 607–613. (Seilacher's proposal for the biological affinity of the Ediacaran Fauna.)

Shen, Y., Zhang, T., Hoffman, P. F., 2008. On the coevolution of Ediacaran oceans and animals. *Proceedings of the National Academy of Sciences of the United States of*

America 105, 7376–7381. (Shows the development of deep-water oxygenation around 580 million years ago.)

Sahoo, S. K., Planavsky, N. J., Kendall, B., Wang, X., Shi, X., Scott, C., Anbar, A. D., et al., 2012. Ocean oxygenation in the wake of the Marinoan glaciation. *Nature* 489, 546–549. (Provides evidence of ocean oxygenation just after the Marinoan Glaciation 630 million years ago.)

Sprigg, R. C., 1947. Early Cambrian (?) jellyfishes from the Flinders ranges, South Australia. *Transactions of the Royal Society of South Australia* 71, 212–224. (The paper that put Ediacaran fossils on the map.)

CHAPTER 11

Bergman, N. M., Lenton, T. M., Watson, A. J., 2004. COPSE: A new model of biogeochemical cycling over Phanerozoic time. *American Journal of Science* 304, 397–437. (Isotope-independent model of oxygen evolution through the Phanerozoic based on a limited number of input "drivers.")

Berner, R. A., 1987. Models for carbon and sulfur cycles and atmospheric oxygen: Application to peleozoic geologic history. *American Journal of Science* 287, 177–196. (Bob Berner's first attempt at an oxygen regulation model. The idea of rapid recycling is introduced.)

Berner, R. A., 2006. GEOCARBSULF: A combined model for Phanerozoic atmospheric O_2 and CO_2. *Geochimica et Cosmochimica Acta* 70, 5653–5664. (A fully coupled oxygen, carbon, and sulfur model with numerous outputs including atmospheric CO_2, O_2, and a variety of chemical constituents of the oceans. Bob Berner's most advanced model platform.)

Berner, R. A., Canfield, D. E., 1989. A model for atmospheric oxygen over Phanerozoic time. *American Journal of Science* 289, 333–361. (Atmospheric oxygen levels from rock adundance data.)

Berner, R. A., Lasaga, A. C., Garrels, R. M., 1983. The carbonate-silicate geochemical cycle and its effect on atmospheric carbon dioxide over the past 100 million years. *American Journal of Science* 283, 641–683. (Classic paper on modeling the history of atmospheric CO_2.)

Butterfield, N. J., 2011. Animals and the invention of the Phanerozoic Earth system. *Trends in Ecology and Evolution* 26, 81–87. (Nick Butterfield's view on the role of animals in shaping their chemical environment.)

Dahl, T. W., Hammarlund, E. U., 2011. Do large predatory fish track ocean oxygenation? *Communicative & Integrative Biology* 4, 1–3. (A nice model study showing how fish metabolic rate and size are coupled to levels of available oxygen.)

Dahl, T. W., Hammarlund, E. U., Anbar, A. D., Bond, D.P.G., Gill, B. C., Gordon, G. W., Knoll, A. H., et al., 2010. Devonian rise in atmospheric oxygen correlated to the radiations of terrestrial plants and large predatory fish. *Proceedings of the National Academy of Sciences of the United States of America* 107, 17911–17915. (Presents molybdenum isotope evidence for a rise in atmospheric oxygen during the Devonian Period.)

Garrels, R. M., Lerman, A., 1981. Phanerozoic cycles of sedimentary carbon and sulfur. *Proceedings of the National Academy of Sciences of the United States of America* 78, 4652–4656. (The paper that started it all. An insightful model of the evolution of the sulfur and carbon cycles through the Phanerozoic.)

Garrels, R. M., Mackenzie, F. T., 1971. *Evolution of Sedimentary Rocks*. W.W. Norton, New York. (A true classic. Wraps the newly found appreciation for plate tectonics with fundamental insights into the composition and geologic cycling of sedimentary rocks.)

Harlé, É., Harlé, A., 1911. Le vol de grands reptiles et insectes disparus semble indiquer une pression atmosphérigue élevée. *Bulletin de la Société Géologique de France* 11, 118–121. (An early, and perhaps the first, discussion of atmospheric oxygen levels and insect gigantism.)

Harrison, J. F., Kaiser, A., VandenBrooks, J. M., 2010. Atmospheric oxygen level and the evolution of insect body size. *Proceedings of the Royal Society of London B* 277, 1937–1946. (A nice review as to how insect size is influenced by oxygen levels in experimental systems.)

Kump, L. R., Garrels, R. M., 1986. Modeling atmospheric O_2 in the global sedimentary redox cycle. *American Journal of Science* 286, 337–360. (The first comprehensive model of the evolution of atmospheric oxygen through geologic time. Based on the modeling philosophy of Garrels and Lerman.)

Lamsdell, J. C., Braddy, S. J., 2010. Cope's Rule and Romer's theory: Patterns of diversity and gigantism in eurypterids and Palaeozoic vertebrates. *Biology Letters* 6, 265–269. (Nice exploration of the history of the evolution of size among euypterids, which links to the evolution of atmospheric oxygen levels.)

Ward, P. D., 2006. *Out of Thin Air*. Joseph Henry Press, Washington, DC. (An exploration of how changes in atmospheric oxygen have influenced animal evolution throughout the Phanerozoic Eon. Based very much on the results of the GEOCARBSULF model.)

CHAPTER 12

Kump, L. R., 2008. The rise of atmospheric oxygen. *Nature* 451, 277–278. (Excellent mini-review on the history of atmospheric oxygen.)

INDEX

Note: Page numbers in italic type indicate illustrations.